Engineering Materials

This series provides topical information on innovative, structural and functional materials and composites with applications in optical, electrical, mechanical, civil, aeronautical, medical, bio- and nano-engineering. The individual volumes are complete, comprehensive monographs covering the structure, properties, manufacturing process and applications of these materials. This multidisciplinary series is devoted to professionals, students and all those interested in the latest developments in the Materials Science field, that look for a carefully selected collection of high quality review articles on their respective field of expertise.

More information about this series at http://www.springer.com/series/4288

Md Rezaur Rahman
Editor

Bamboo Polymer Nanocomposites

Preparation for Sustainable Applications

Springer

Editor
Md Rezaur Rahman
Faculty of Engineering
Universiti Malaysia Sarawak
Kota Samarahan, Sarawak, Malaysia

ISSN 1612-1317 ISSN 1868-1212 (electronic)
Engineering Materials
ISBN 978-3-030-68089-3 ISBN 978-3-030-68090-9 (eBook)
https://doi.org/10.1007/978-3-030-68090-9

This Springer imprint is published by the registered company Springer Nature Switzerland AG
The registered company address is: Gewerbestrasse 11, 6330 Cham, Switzerland

Contents

Introduction of Various Types of Bamboo Species and Its Nanocomposites Preparation

Muhammad Khusairy Bin Bakri, Md Rezaur Rahman, and Muhammad Adamu

Abstract This chapter explores the bamboo species and possibility of its nanocomposites, which is directly correlated with the desirable physical and mechanical performance. The distribution of bamboo and the anatomy of bamboo was reported based on the region. The chemical composition of bamboo and its extraction process were described in this report. The preparation of nanocomposites in the present of montmorillonite was also investigated.

Keywords Bamboo · Species · Nanocomposites · Potentiality · Commercial

1 Natural Fiber

Natural fibers are categorized depending on various selection criteria's and sources [1]. The bamboo plant can grow as big as giant size depending on the soil, moisture, habitat's, regions, temperature, sunlight, and tolerance, while its woody stem may have various characteristics such as fast-growing like grasses [2, 3]. Furthermore, most researchers show that there are many high-quality fiber and cellulose extracted from bamboos during the pulping process [4, 5]. In Asia, Middle and South America, major advantage using bamboo was due to its abundantly available fiber and main natural resource in the regions [6]. Due to its high strength to weight ratio, bamboo fibers are frequently referred as natural glass fibers. The bamboo fiber is comparable to the steel strength, whereas it was reported that the tensile strength may reach up to 370 MPa on tensile loading applications [7]. In addition, bamboo fibers were widely used in composite and construction industries due to its highly developed potential purposeful tool [8, 9].

The diverse matrices usage of bamboo-based have boosted the new eco-friendly and cost-effective bio-composites market values on the composites production [10].

M. K. B. Bakri (✉) · M. R. Rahman · M. Adamu
Faculty of Engineering, Universiti Malaysia Sarawak, Jalan Datuk Mohammad Musa, 94300, Kota Samarahan, Sarawak, Malaysia

M. R. Rahman
e-mail: rmrezaur@unimas.my

M. R. Rahman (ed.), *Bamboo Polymer Nanocomposites*, Engineering Materials,
https://doi.org/10.1007/978-3-030-68090-9_1

Different from its traditional counterparts, the future of sustainable bamboo-based composite industry was accountable to help on the utilization of bamboo to be more profitable in many applications. The testing and analysis for the future effective and high-quality bamboo fiber characterization and development on its composites were done to ensure it is aligned with the international standards [11–13]. The fibers extracted from plants must meet the composite industry requirements, in such depend directly on the fiber extraction methods [14, 15]. Competitively, different methods create different structures and varieties, which also influence by the mechanical, chemical and thermal properties of the potential bamboo in the polymer composite industry [14, 15]. The selected materials for matrix used on the bamboo fibers to fabricate the bio-composite has brought huge improvements on its properties, especially when the new natural fiber treatment and development processes, methods, and technologies was applied [16–19]. The cost competitiveness acceptance also improved, which was correlated directly with the desirable physical and mechanical performance of production facilitation [16–19].

2 Bamboo

In climatic torrential rain condition, a perennial bamboo plant can grow up to 40 m in height. Widely in many applications, bamboo can be used as carpentry, construction industries, food, weaving, plaiting and tools [20–23]. Compared with other various products and materials, curtains produced bamboo fibers were observed bamboo to absorb a certain electromagnetic radiation that at various wavelengths, i.e. x-ray, infrared, etc., which was less harmful to the human body [24]. Around the world, especially in tropical and sub-tropical regions, tradition of using bamboo was well-established, especially related and incorporated with building material [25, 26]. In addition, there were estimated around 1200 bamboo species grown in Asia, Latin America and Africa regions [27–29]. Due to bamboo strong and versatile properties, either the stick, pole, fiber, or cellulose is used in construction tools and applications, i.e. roofing, flooring, scaffolding, walls, piping, and concrete reinforcement [30, 31]. Bamboo also is usually used as a decoration element. In addition, and with consideration of earthquake-prone area, the extreme lightweight bamboo is structurally useful to be used as it have greater absorption capacity [32, 33].

Recent studies also showed for the eco-composites, the composites environmentally friendly fabrication, especially the extracted raw bamboo fiber from steam explosion technique had advantageous mechanical properties [34]. As compared with conventional glass fibers, bamboo fibers had enough specific strength [7]. Respectively, due the impregnation and reductions of voids numbers, there was 30% increase in modulus and 15% increase tensile strength, especially those made from mechanically extracted fibers [35, 36].

Table 1 Bamboo coverage based on regions and countries

Bamboo regions	Countries	Percentages
Asia–Pacific Region	Bangladesh, Burma, Cambodia, China, India, Indonesia, Japan, Korea, Malaysia, Philippines, Sri Lanka and Thailand	65
American Region (Latin America, South America and North America)	Brazil, Costa Rica, Columbia, Guatemala, Honduras, Mexico, Nicaragua, Venezuela and some European countries	28
African Region	Eastern Sudan and Mozambique	7

3 Bamboo Distribution

Bamboo mostly grown in Asia–Pacific, Africa, Europe, America and North America continents [37, 38]. The Asia–Pacific continent was known to have the largest bamboo reserve. As shown in Table 1, bamboo was identified differently depending on their respective regions, i.e. in Vietnam as "the brother", in India as "wood of the poor" in India, and in China as "friend of the people". China, India, Indonesia, Myanmar, and Vietnam are among the Asian countries that reserved a large bamboo plantation land [39]. As part of global dominance, China has created an extensive awareness, especially on its existing bamboo plantations, whereas it is observed that the sympodial type of bamboo, which was a noteworthily increase around 30% compared with the monopodial bamboo [40]. Contrary with other Asia region, another part of regions such as South America widely unused its bamboo, this abundance drives it to grow abundantly and naturally without the need for cultivation, which made it a better choice of fibers over other natural fibers.

In many countries, the scarcity of forest resources, i.e. bamboo as crops are being utilized, especially for the used in composite reinforcements, which is also listed in the topmost crops [41]. Bamboo is not fully functionality explored in many parts of Asia and South America despite of its abundance, even though most of the Asian countries considered it as a natural engineering material [42]. Bamboo plant would need about six to eight months to be mature in size, depends on its species and variations [43]. Bamboo was often referred to as weed due to its fast growth and widespread. Even though it is not vastly used due to its unwavering claim of resourceable sustainable plant, Bamboo can also be utilized in small-scale home décor and construction [36].

4 Bamboo Anatomy

From the outside view, the bamboo culm appears in a ring-like form, which was synonymously in diaphragm hollow cylinder shape, whereas the ring divides every inner side culm of the internode space between the two rings from the grow branches

[44]. Thus, each node distance is different between varied species of bamboo. While in the outer part of the strengthens bamboo culm, it consists of few vascular bundles with special internodes [42, 45]. This antinodes size depends directly with the culm height. For bamboo characterization, most of it depends on the average size, density, and number of vascular bundle parameters [46]. It is usually found in the bamboo culm, whereas the parameters may be different depending on the bamboo species [46]. Therefore, the bamboo usability was determined by its culm anatomy, which also reflected by its parameters, which later define its physical properties. Generally, smaller bamboo diameter has higher fiber density. Bamboo base section can withstand a larger force compared to the upper section of the bamboo, even though it has lower strength [47]. This was due to the bamboo hollow tube, whereas from the inside hollow structure acts as an internal pressure to support the bamboo structure.

Bamboo culm is a microstructure. It is well transversely distributed around its thick wall, rooted in the parenchyma fleshy tissue, and comes with numerous vascular bundle [48]. All of the nodes to the culm are connected, whereas the vascular bundles and its sheaths reinforce the bamboo culm. Even though it is less appeared in the inner side, these vascular bundles were surrounded by fibrils (sclerenchyma cells), which were deeply scattered along the outer culm wall side [48]. From the base and top culm, the bamboo size and density are different, and the vascular bundles contain mostly xylem and phloem cells [49].

Xylem functionality is to transfers water, while the phloem transports sugars and nutrients to all the plant parts. Each of the vascular bundle contains pentagonal and hexagonal shapes, and elementary fiber strand. Nano-fibrils usually embedded and bound together with the hemicellulose and lignin, which allied rigorously in the vascular bundle [50]. Figures 1 and 2 show the bamboo culm and fiber structures [51]. To understand bamboo anatomical structure better, several studies on its microstructure need to be carried out. Most of the bamboo was made up of 60% cellulose and 32% lignin [52].

5 Bamboo Chemical Compositions

Most studies show that the bamboos age, region, and environmental conditions, determine the fiber structures, which sorted the percentage composition different in the cellulose, hemicellulose, lignin, and ash. Almost 90% of the bamboo total weight were fiber made, while the 10% remaining constituents are ash, fat, pectin, pigments, protein, and tannins [53]. Many of these constituents are stationed in the special organelles and cell cavity, which are physiological vital to determinants the bamboo activity. Figure 3 shows the bamboo fiber chemical constituents.

A lignin component was difficult to be removed, as it is resistance to most of various alkaline based materials. Lignin also provided bamboo with stiffness and its yellowish color properties, while other non-cellulosic component, i.e. ash, pectin, etc. also provides the density, strength, flexibility, and moisture. In addition, the

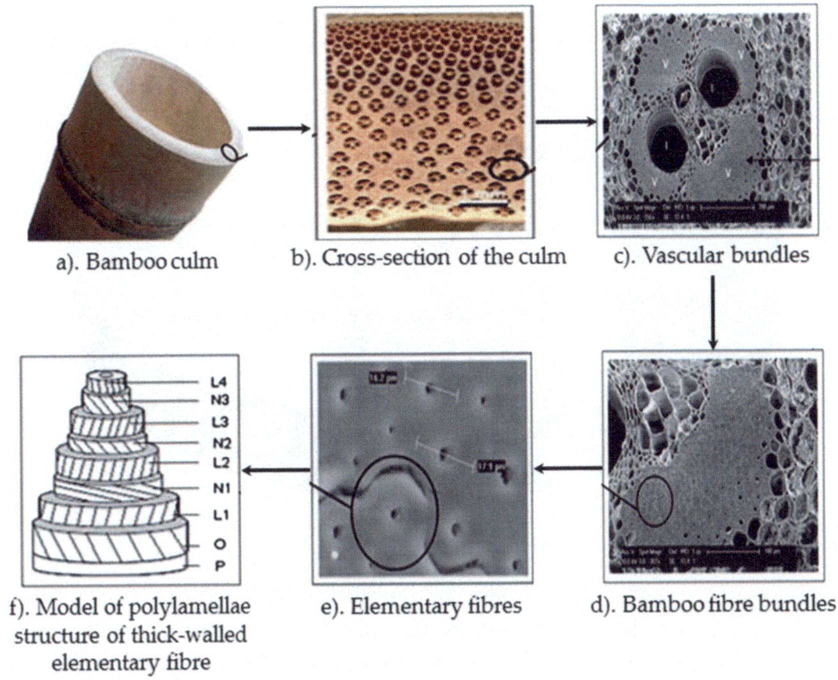

Fig. 1 Bamboo **a** culm, **b** culm cross-section, **c** vascular bundles, **d** fiber strand, **e** elementary fibers, and **f** poly lamellae structure model [51]

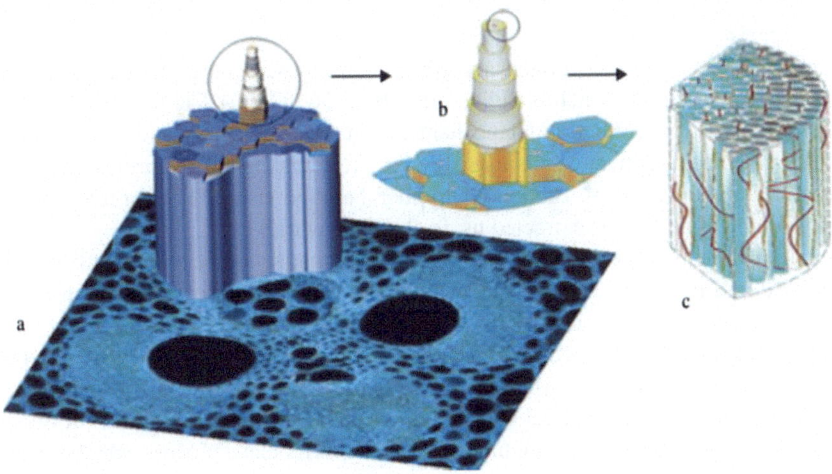

Fig. 2 Bamboo **a** vascular bundle, **b** 10–20 μm elementary fiber, and **c** 1–10 μm nanofibril of lignin and hemicellulose [51]

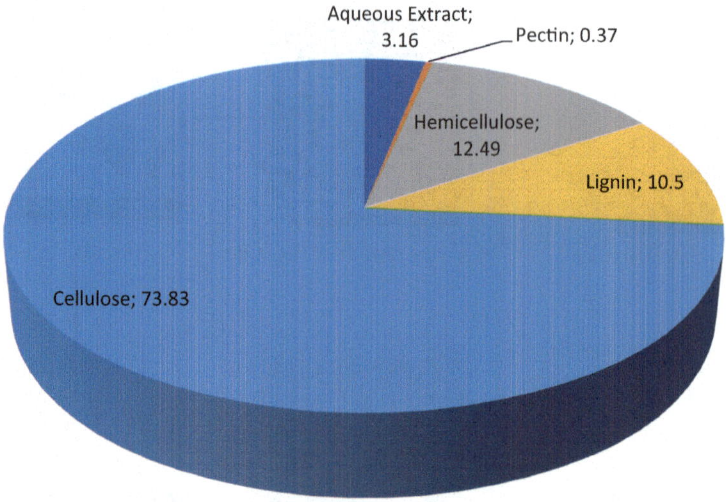

Fig. 3 Bamboo fiber chemical constituents [54]

composition comparisons with other natural fibers shows that the bamboo fibers mechanical properties have an almost similar source of cellulose and lignin [54, 55].

6 Bamboo Fiber Extraction

Bamboo fibers were categorized according to the process and methods used for during the extraction, which is divided into two [56, 57]: (i) The original bamboo fiber is directly extracted through mechanical and physical means, without the usage of any chemical additive, which also called as "raw", "original", "pure bamboo" or "natural bamboo", and (ii) Bamboo pulp fiber, bamboo viscose fiber or regenerated cellulose bamboo fiber is extracted using chemical additives.

Initially, the bamboo strips were split into many parts and the bamboo fiber were obtained and processes through mechanical and chemical processes. This process also depends on the repeated number of times bamboo fibers were used and how much the load was subjected too. The chemical process was done by yielding the cellulose fibers with alkali hydrolysis of sodium hydroxide (NaOH) before multi-phase bleaching was carried out via passing through carbon disulphide for the alkali treated cellulose fibers. Most manufacturers used this repeated process, which is least time-consuming bamboo fibers procedure [58].

The mechanical process involved typical initial crushed bamboo process without enzymatic treatment. This led to the spongy mass and individual fibers formation, which were obtained with the mechanical comb aid. This method was more environment-friendly compared to the chemical process, even though it is less economical [59]. It was reported that it is more convenient extract rough or fine bamboo and divide the fiber depending on the preparation method [60].

The rough bamboo fibers can be prepared by cutting, boiling and separation. While to obtained fine bamboo, the bamboo was fermented with enzymes and boiled. It also can be washed, bleached with acid treatment and finally soaked in oil and air-dried. However, at the end, by compared these two fiber extractions through mechanical process over chemical process, the major advantage is that it creates better environmental characteristic to the extracted fiber [61].

7 Mechanical Extraction Methods

For special composite reinforcement applications, especially to extract the fiber through mechanical means, this method may involve many different processes such as steam explosion, retting, crushing, grinding and rolling of the bamboo in a mill.

7.1 Steam Explosion

Steam explosion is commonly used methods to produce pulp. It also consumes less process energy, while also involves in the separation cell walls of bamboo plant. This method has been since 1962, which was mainly utilized and used in the pulp industry. Even though the resulting fibers from the process are dark in color and rigid, it was also appropriated to be used to separate lignin from the plant surface [62]. Unfortunately, it was also impossible to completely remove lignin from the bamboo fiber through the steam explosion process. However, bamboo fiber cotton (BFC) can be made through machine mixing, whereas the lignin was separated as a remnant from the fibers [63]. It was also observed that the tensile strength of BFC reinforced composite had much greater weight percentage compared with those containing only bamboo fiber. The bamboo was cut and heated in an autoclave at 175 °C and 0.7–0.8 MPa for about 60 min. The fibers then were washed with soap in hot water at 90–95 °C. For several times, the same process was repeated to ensure and prevent the cell walls from fracture, before dried it in an oven at 105 °C for 24 h.

The steam was instantly released in 5 min to carry out the adhesion reduction between the extracted fibers and resin [64, 65]. The process was repeated to ensure the ash was removed and the lignin content remnant was condensed on the fiber surface [64, 65]. To enabling fiber extraction during the steam explosion process, the bamboo fiber was softened by cracking of the fiber cell wall. The partial lignin decomposition happened during the crushing of the softest bamboo fiber surface cell wall, which reduces the shear resistance [64]. This lignin was ultrasonically washed from the fibers, while treat with isocyanate to remove unexpended cells. The results showed that the extracted fiber after the steam explosion had high tensile strength as compared to isocyanate treated fibers. The reason behind this was that the chemically treated fiber created a very weak interface between the soft cells and its fibers, which reduced greater tensile strength extent produced by fiber reinforced thermoplastic composites [66]. Therefore, newly improved or different treatment technique is required to enhance the adhesion between the bamboo fiber and matrix [67].

7.2 Retting

Retting process or procedure involves the cylindrical peeling section of the culm to obtain its strips. The bamboo bark was removed before the peeling process. After the strips were parked in bundles and wetted with water for at least three days, it was compressed and trimmed with a sharp-edged knife. During the scraping process, the fiber quality may be affected, especially along the fiber's length, which the fiber can be broken due to the uneven surface done due to the scraping process [68]. Another study shows that without scraping or combing the bamboo, the raw bamboo without removal of the bamboo epidermis and node can be cut into few longitudinal fragments [69]. Aerobic and anaerobic retting can be done to separate the bundles from the culm effectively. The bamboo strips were rinsed with water, while the culms were fermented in water under room temperature for two months. It also showed that the extracted fiber bundle consisted of a single invariably length fiber [70].

7.3 Crushing

The raw bamboo was extracted using the roller crusher, to cut it into small pieces, before extracting the coarse fiber using a pin-roller during the extraction process. The coarse bamboo fiber was boiled at 90 °C for 10 h in a dehydrator to remove the fat before drying it in a rotary dryer [71]. This challenge process produces short fibers or sometimes powder due to over mechanical process [72].

7.4 Grinding

Grinding involves the bamboo culm cutting without nodes into a few strips before soaking in water for 24 h [73]. With a sharp knife, the wetted strips were manually cut into smaller pieces. The longer bamboo strips were cut into small chips, while an extruder was used to obtain a wide strip. A high-speed blender was used to ensure that the bamboo fibers were in smaller sizes. It was then separated with sieves into many sizes and apertures. An oven was used to dry the extracted fibers at 105 °C for 72 h. A high tensile load was used to obtain long fibers, which may increase the transverse length [74]. In the dried bamboo strand production, this grinding system has been utilized, which can be used to study nano clay as well [75].

7.5 Rolling Mill

The rolling mill method involves the bamboo culm cutting from the nodes into smaller pieces. The strips usually cut into 1 mm thickness, while maintaining its low speed and pressure. The strips were soaked in water for an hour to allow fibers separation. The rolled strips obtained were soaked in water for 30 min before slicing it into small pieces. After sun dried for a maximum of 2 weeks, around 220–270 mm fibers size were obtained [49]. It was also reported that without necessarily soaking the bamboo strips in water, the fiber can be extracted and obtained thru the compression between the bamboo and two pairs of steel cylinders [45]. Furthermore, the soaking of bamboo in water is soften the lignin and ease the slicing process which allowed the fibers passed through the roller easily. However, this will bring much about the reduction in their bonding strength. Typically, 30–60 cm in length of fibers can be extracted from this process [76].

7.6 Chemical

Alkali or acid chemical retting procedure was used to reduce or remove elementary lignin content in the fibers, whereas the chemical assisted natural, or degumming. However, this treatment may influence other bamboo microstructure components, i.e. cellulose, hemicellulose and pectin [77].

7.7 Simultaneous Extraction and Degumming

Few researchers used a combination of chemicals and enzymes to extract fine and soft fiber before the degumming process. This process was achieved by removing the content of pectin and gums from the decorticated bamboo strips [78]. The enzyme would degrade and produces a very good separation of the cellulose fibers of the gum material, which located in the middle lamella and the cell wall [79].

7.8 Alkaline or Acid Retting

The processing application method involves bamboo stripping and heating in a stainless-steel vessel containing 1.5 M NaOH solution at 70 °C for 5 h. After the heat treatment, to separate the fibers of alkaline treated bamboo strips, the steel pressing machine was used [74, 80]. It was noted that the extraction process creates less fiber damage. It was reported that the bamboo was sieved into smaller chips and soaked in 1 M NaOH for at 70 °C 2 h. To further process the enhances bamboo fiber separation, the cellulose and nanocellulose stimulation needed to be done [70]. This procedure was repeated for several times, under controlled pressure, which further allow the fiber extraction in the pulp form. However, the disadvantage of this extraction method as the extraction continues is the development of larger fiber bundles or materialization.

In a separate study, the fiber was extracted with trifluoroacetic acid (TAA) solutions, and the bamboo strips were soaked in 1 M NaOH for 72 h. It was found that the lignin dissolve completely in both NaOH and TAA [81]. When soaked in NaOH, the remaining lignin in the middle lamella was substantially removed by TTA [16]. Hence, alkaline solutions offer great interfacial bonding for composites between fiber and matrix compared to other methods such as degumming or mechanical extraction.

7.9 Chemical Retting

Chemically assist the retting process was also used to remove water and lignin from bamboo fibers. In longitudinal section, the bamboo culm was sliced into very thin slabs or chunks, after the fibers were separated manually. It was immersed in 1–3% of $Zn(NO_3)_2$ solution, while the maintained the liquor to a bamboo ratio of 1:20. After boiling it in water for 1 h, it was kept at 40 °C for 16 h with neutral pH. Compared to alkaline or acid retting process, this enhanced process efficiently removes lignin. It was also observed that the bamboo fiber moisture content was very high. Trujillo and López [82] shows the process by using the 2 cm sliced bamboo culm chips was roasted at 150 °C for 30 min. The chips were soaked in water for 24 h at 60 °C, dried in the air, and rolled on a flat surface to further remove its impurities [82]. This procedure was repeated, while the afterward fiber bundles were soaked and cooked in 0.5% of NaOH, 2% of $Na_5P_3O_{10}$, 2% of Na_2SO_3 and 2% of NaSi solutions for 60 min, at 60 °C [82]. The bamboo to liquor ratio is maintained at 1:20. Before the bamboo fiber was washed in hot water, the acid treatment with 0.5% diethylenetriamine pent-acetic acid and 0.04% xylanase acid was done for 60 min at 70 °C and pH 6.5. The bamboo fibers were recooked at 100 °C for 60 min using the same procedure with only 0.7% of NaOH is used. The bamboo fibers were bleached and repeatedly done with 0.2% of NaOH, 4% of H_2O_2, and 0.5% of NaSi in a polyethylene bag for another 50 min [83]. Before it was emulsified for five days, the bamboo fibers were refined with 0.5% of $H2SO_4$ acid for 10 min. It can be concluded that the bamboo fiber obtained was very small orientation in angle. While as compared to fibers obtained from cotton, flax and ramie, bamboo exterior macro fibrils would serve as reinforcement with good property [84].

7.10 Combined Mechanical and Chemical Extraction

This method involves the mechanical extraction combinations with chemical extraction. By using the roller molding method (RMM) and compression molding methods (CMM), the mechanical extraction can be achieved. The chemical extraction was done using alkaline and acid solutions after mechanical process [85]. In a separate study, after a 10-ton load placed at both bamboo ends strips to pull and extract the fibers, CMM was used to pressurize the soaked bamboo strips in a bed of alkaline solutions. To obtain high good quality fibers, two key factors are essential: (i) bed thickness, and (ii) compression time [42]. RMM method usually used to flatten the bamboo strips by involving the use of two rollers at fixed both ends, and/or any one end can be made to rotate, while the other end is fixed. The chemical and mechanical application processes would enhance quick strips separation into various bamboo fiber sizes. For the use of RMM, two factors are essential, which is the size of compression mold mounts and the diameter of the rollers, whereas these two determine the fiber amount which can be extracted [45].

In another study, a roller was utilized in extracting the fibers, whereas the nodes of the bamboo culm were detached, while internodes were sliced along the longitudinal direction using slicer. The bamboo strips were then soaked in 1, 2, and 3 concentrations of NaOH solution at 70 °C for 10 h. The fibers soaked in 1% of NaOH mechanical properties were found to be higher than those fibers immersed in other concentrations. Using a roller looser the alkali-treated fiber strips, and extracted the small fibers, which later were dried in an oven at 105 °C for 24 h [86].

8 Bamboo Nanocrystals Extraction for Nanocomposites

Bamboo contained large source of cellulose and fiber. It was valuable in the biocomposites and nano-scale particles production because the fibers can be made able into natural nanoscale components [87]. In addition, cellulose fibers are cheap, environment-friendly, and easily found from plant fibers. Thus, the natural cellulose fibers suitable for preparation of nanocomposites. Researchers also reported that the nanofibers and nanocrystals can be extracted from many plants [88]. There are many various methods can be used to nanofibers and nanocrystals from natural materials, but the foremost approach to prepare cellulose nanofibers and nanocrystals, is to embroil it using mechanical treatment, enzymatic treatment, and chemical modification [89]. However, the cellulose structure stability and chemical reagents blockage from reacting with active fiber groups, make it quite uneasy to obtain cellulose nanofibers and nanocrystals. Renowned mechanical treatments, i.e. grinding, a high-pressure homogenizer, and ultra-sonication were utilized to facilitate the chemical process [54].

9 Nanocomposites Fabrication

9.1 Montmorillonite for Nanocomposites

Few researchers reported that natural fiber choice depends on its biodegradable, recyclable, and sustainable potential [90, 91]. The final composite characteristics depend on the inorganic filler nature, microstructure, and compositions as well as the interfacial interactions of the microstructure of the natural fibers [7, 92]. Sanal [22] reported that the consumer desirability on bamboo as a choice was established, which eased by its high-end sustainable quality industrial products, and the high strength-weight ratio that makes it arguably an attractive alternative to steel in tensile loading applications. In engineering material comparison between conventional macro and micro composites, the nanocomposites significant shown improvement in its mechanical, thermal, physical as well as decreasing component weight required [93, 94].

The suitable selection of effective interactions an inorganic filler significantly changed the microstructure, which was vital in the nanocomposites materials production as shown by Chen et al. [95]. Studies carried out by Reddy [96] and Essabir et al. [97] shown that clay was a suitable material to be used as a nanofiller. Silva et al. [98] reported that nanofiller clay significantly has better physical and mechanical properties compared to the conventional polymer's filler. In addition to its high content of alkaline earth metals, i.e. magnesium iron, the choice of montmorillonite as nanofiller among various types of clay was a result of its swelling capacity, high surface-volume ratio, strong absorption/adsorption properties [99]. Additionally, nanofillers clay used in the polymer industry was rooted due to its lightness, and mechanical properties of natural fibers. However, Manalo et al. [80] reported that the major bamboo fiber challenge resulted the poor mechanical properties of the composites, which was due to the inherent flaws within fibers that reduced their compatibility with polymer matrices.

9.2 Polymers in Nanocomposite

Poly (vinyl alcohol) (PVA) is known to have good compatibilities and property for the fabrication of composites. It was known to be water-soluble synthetic polymer, biocompatible and biodegradable active, and semi crystalline structure [100]. In combination with the cellulose or nanocellulose, a strong hydrogen bond result shows that PVA has a very good ability to interact with most hydrophilic materials to produce nanocomposite [101]. In composite, the polymers react with the hydroxyl groups to disperse and enhanced interfacial adhesion by chemical bonding, and subsequently improved the mechanical properties [102]. Moreover, by adding bamboo nanofibrils into PVA, Mousa et al. [103], Guimarães et al. [11] and Tang et al. [104] reported good compatibility whereby increased its mechanical properties. Another approach to overcome the poor barrier properties and low mechanical resistance was by using

NaOH onto the fibers, which create new forms of nanocomposites [48, 58, 95, 105–107]. Recent findings by Khan et al. [48] exposed a composite yield and optimized the reinforcing materials by using NaOH to attain high tensile strength.

Researchers such as Islam et al. [108], Daneshpayeh et al. [109], Olivera et al. [110] and Zhu et al. [111] reported that elastomers are useful due to their broad properties range. The polymeric structure composed of long and flexible chains, which enable it to sustain very large deformations [110]. Generally, a very complex mechanical viscoelastic elastomer has good response and feedback [93]. Styrene application as reinforcing filler was found to increase the mechanical properties (modulus and strength) in composites [48].

10 Summary

In this Chapter, the used of bamboo fibers with propose matrix materials to be used as nanocomposites has brought an improvement, especially in the treatment of extracting fibers from bamboo which is using high processing technologies for nanocomposites. This improves the cost competitiveness acceptance, which is correlated directly with the facilitation of production with desirable physical and mechanical performance. The bamboo nanocomposites were utilized in small-scale construction and home décor, even though it has not seen the light of vast usage, especially with its unwavering claim of resource-able sustainable plant. In addition, the suitable inorganic filler introduces the effective interactions between polymer and fiber, which significantly changed the microstructure, of nanocomposite.

Acknowledgements The authors would like to acknowledge Universiti Malaysia Sarawak (UNIMAS) for the support.

References

1. Roslan, S.A.H., Rasid, Z.A., Hassan, M.Z.: The natural fiber composites based on bamboo fibers: a review. ARPN J. Eng. Appl. Sci. **10**(15), 6279–6288 (2015). https://www.arpnjournals.com/jeas/research_papers/rp_2015/jeas_0815_2402.pdf
2. Rajan, K.P., Veena, N.R., Maria, H.J., Rajan, R., Skrifvars, M., Joseph, K.: Extraction of bamboo microfibrils and development of biocomposites based on polyhydroxybutyrate and bamboo microfibrils. J. Compos. Mater. **45**(12), 1325–1329 (2011). https://doi.org/10.1177/0021998310381543
3. Peng, Z., Lu, Y., Li, L., Zhao, Q., Feng, Q., Gao, Z., Lu, H., Hu, T., Yao, N., Liu, K., Li, Y., Fan, D., Guo, Y., Li, W., Lu, Y., Weng, Q., Zhou, C., Zhang, L., Huang, T., Jiang, Z.: The draft genome of the fast-growing non-timber forest species moso bamboo (Phyllostachys Heterocycla). Nat. Genet. **45**(4), 456–461 (2013). https://doi.org/10.1038/ng.2569
4. Sharma, B., Shah, D.U., Beaugrand, J., Janecek, E.-R., Scherman, O.A., Ramage, M.H.: Chemical composition of processed bamboo for structural applications. Cellulose **25**(1), 3255–3266 (2018). https://doi.org/10.1007/s10570-018-1789-0

5. Lu, T., Liu, S., Jiang, M., Xu, X., Wang, Y., Wang, Z., Gou, J., Hui, D., Zhou, Z.: Effects of modifications of bamboo cellulose fibers on the improved mechanical properties of cellulose reinforced poly(lactic acid) composites. Compos. B Eng. **62**(1), 191–197 (2014). https://doi.org/10.1016/j.compositesb.2014.02.030
6. Khalil, H.A., Alwani, M.S., Islam, M.N., Suhaily, S.S., Dungani, R., H'ng, Y.M., Jawaid, M.: The use of bamboo fibres as reinforcements in composites. Biofiber Reinf Compos Mater **12**(2), 488–524 (2015). https://doi.org/10.1533/9781782421276.4.488
7. Yu, Y., Huang, X., Yu, W.: A novel process to improve yield and mechanical performance of bamboo fiber reinforced composite via mechanical treatments. Compos. B: Eng. **56**(8), 48–53 (2014). https://doi.org/10.1016/j.compositeb.2013.08.007
8. Sharma, A.K., Chaudhary, G., Kaushal, I., Bhardwaj, U., Mishra, A.: Studies on nanocomposites of polyaniline using different substrates. Am. J. Polym. Sci. **5**(1), 1–6 (2015). https://doi.org/10.5923/s.ajps.201501.01
9. Sharma, B., Gatóo, A., Bock, M., Ramage, M.: Engineered bamboo for structural applications. Constr. Build. Mater. **81**(23), 66–73 (2015). https://doi.org/10.1016/j.conbuildmat.2015.01.077
10. Akinlabi, E.T., Anane-Fenin, K., Akwada, D.R.: Bamboo taxonomy and distribution across the globe. Bamboo **9**(1), 37–49 (2017). https://doi.org/10.1007/978-3-319-56808-9_1
11. Guimarães, M., Botaro, V.R., Novack, K.M., Teixeira, F.G., Tonoli, G.H.D.: Starch/PVA-based nanocomposites reinforced with bamboo nanofibrils. Ind. Crops Prod. **14**(3), 1221–1243 (2015). https://doi.org/10.1016/j.indcrop.2015.03.014
12. Latif, S.S., Nahar, S., Hasan, M.: Fabrication and electrical characterization of bamboo fiber-reinforced polypropylene composite. J. Reinf. Plast. Compos. **34**(3), 187–195 (2015). https://doi.org/10.1177/0731684414565941
13. Palombini, F.L., Kindlein, W., de Oliveira, B.F., de Araujo Mariath, J.E.: Bionics and design: 3D microstructural characterization and numerical analysis of bamboo based on X-ray microtomography. Mater. Charact. **120**(45), 357–368 (2016). https://doi.org/10.1016/j.matchar.2016.09.022
14. Mounika, M., Ramaniah, K., Ratna Prasad, A.V., Rao, K.M., Hema Chandra Reddy, K.: Thermal conductivity characterization of bamboo fiber reinforced polyester composite. J. Mater. Environ. Sci. **3**(6), 1109–1116 (2012). https://www.jmaterenvironsci.com/Document/vol3/vol3_N6/112-JMES-322-2011-Mounika.pdf
15. Takagi, H., Fujii, T.: Mechanical characterization of bamboo fiber-reinforced green composites. Key Eng. Mater. **557–578**(1), 81–84 (2013). https://doi.org/10.4028/www.scientific.net/KEM.577-578.81
16. Liu, D.G., Song, J.W., Anderson, D.P., Chang, P.R., Hua, Y.: Bamboo fiber and its reinforced composites: structure and properties. Cellul. **19**(5), 1449–1480 (2012). https://doi.org/10.1007/s10570-012-9741-1
17. Singh, T.J., Samanta, S.: Characterization of natural fiber reinforced composites-bamboo and sisal: a review. Int. J. Res. Eng. Technol. **03**(07), 187–195 (2014)
18. Eberts, W., Siniawski, M.T., Burdiak, T., Polito, N.: Mechanical characterization of bamboo and glass fiber biocomposite laminates. J. Renew. Mater. **3**(4), 259–267 (2015). https://doi.org/10.7569/JRM.2014.634137
19. Thakur, V.K., Kessler, M.: Green biorenewable biocomposites: from knowledge to industrial applications. Green Biorenewable Biocomposite **12**(1), 80–92 (2015). https://doi.org/10.1201/b18092
20. Clark, L.G., Londono, X., Ruiz-Sanchez, E.: Bamboo taxonomy and habitat. In: Bamboo: The Plant and Its Uses, 2(1), 1–30 (2015). https://doi.org/10.1007/978-3-319-14133-6
21. Gohil, P.P., Patel, K., Chaudhary, V., Ramjiyani, R.: Effect of bamboo hybridization and staking sequence on mechanical behavior of bamboo-glass hybrid composite. In: Green Approaches to Biocomposite Materials Science and Engineering, 24(4), 321–340 (2016). https://doi.org/10.4018/978-1-5225-0424-5.ch004.
22. Sanal, I. (2016). Bamboo fiber-reinforced composites. In: Green Approaches to Biocomposite Materials Science and Engineering, 4(5), 987–992. https://doi.org/10.4018/978-1-5225-0424-5.ch011

23. Chongtham, N., Bisht, M.S., Haorongbam, S.: Nutritional properties of bamboo shoots: potential and prospects for utilization as a health food. Compr. Rev. Food Sci. Food Saf. **10**(3), 153–168 (2011). https://doi.org/10.1111/j.1541-4337.2011.00147.x
24. Li, M.F., Sun, S.N., Xu, F., Sun, R.C.: Microwave-assisted organic acid extraction of lignin from bamboo: structure and antioxidant activity investigation. Food Chem. **134**(3), 1392–1398 (2012). https://doi.org/10.1016/j.foodchem.2012.03.037
25. Shen, L., Yang, J., Zhang, R., Shao, C., Song, X.: The benefits and barriers for promoting bamboo as a green building material in China—an integrative analysis. Sustainability **11**(9), 2493 (2019). https://doi.org/10.3390/su11092493
26. Manandhar, R., Kim, J.-H., Kim, J.-T.: Environmental, social and economic sustainability of bamboo and bamboo-based construction materials in buildings. J. Asian Arch. Build. Eng. **18**(2), 49–59 (2019). https://doi.org/10.1080/13467581.2019.1595629
27. Yeasmin, L., Ali, M.N., Gantait, S., Chakraborty, S.: Bamboo: an overview on its genetic diversity and characterization. 3 Biotech **5**(1), 1–11 (2015). https://doi.org/10.1007/s13205-014-0201-5
28. Owen, A.: Bamboo!! Improving island economy and resilience with Guam College students. J. Mar. Isl. Cult. **4**(2), 65–75 (2015). https://doi.org/10.1016/j.imic.2015.09.002
29. Canavan, S., Richardson, D.M., Visser, V., Le Roux, J.J., Vorontsova, M.S., Wilson, J.R.U.: The global distribution of bamboos: assessing correlates of introduction and invasion. AoB PLANTS **9**(1), 1–18 (2017). https://doi.org/10.1093/aobpla/plw078
30. Habibi, S.: Design concepts for the integration of bamboo in contemporary vernacular architecture. Arch. Eng. Des. Manag. **15**(6), 475–489 (2019). https://doi.org/10.1080/17452007.2019.1656596
31. Nurdiah, E.A.: The potential of bamboo as building material in organic shaped buildings. Procedia - Soc. Behav. Sci. **216**(1), 30–38 (2016). https://doi.org/10.1016/j.sbspro.2015.12.004
32. Wang, C., Sarhosis, V., Nikitas, N.: Strengthening/retrofitting techniques on unreinforced masonry structure/element subjected to seismic loads: a literature review. Open Constr. Build. Technol. J. **12**(1), 251–268 (2018). https://doi.org/10.2174/1874836801812010251
33. Xu, Q., Chen, X., Chen, J.-F., Harries, K.A., Chen, L., Wang, Z.: Seismic strengthening of masonry walls using bamboo components. Adv. Struct. Eng. **22**(14), 2982–2997 (2019). https://doi.org/10.1177/1369433219855902
34. Sasaki, C., Wanaka, M., Takagi, H., Tamura, S., Asada, C., Nakamura, Y.: Evaluation of epoxy resins synthesized from steam-exploded bamboo lignin. Ind. Crops Prod. **43**(1), 757–761 (2013). https://doi.org/10.1016/j.indcrop.2012.08.018
35. Hojo, T., Zhilan, X.U., Yang, Y., Hamada, H.: Tensile properties of bamboo, jute and kenaf mat-reinforced composite. Energy Procedia **56**(3), 72–79 (2014). https://doi.org/10.1016/j.egypro.2014.07.133
36. Akinlabi, E.T., Anane-Fenin, K., Akwada, D.R.: Properties of bamboo. Bamboo **9**(3), 131–147 (2017). https://doi.org/10.1007/978-3-319-56808-9_3
37. Canavan S, Wilson J., Richardson D.: Understanding the risks of an emerging global market for cultivating bamboo: considerations for a more responsible dissemination of alien bamboos. 10th World Bamboo Congress, 124–143 (2015)
38. Liu, X., Smith, G.D., Jiang, Z., Bock, M.C.D., Boeck, F., Frith, O., Gatóo, A., Liu, K., Mulligan, H., Semple, K.E., Sharma, B., Ramage, M.: Nomenclature for engineered bamboo. BioResources **11**(1), 1141–1161 (2016). https://doi.org/10.15376/biores.11.1.1141-1161
39. Okubo, K., Fujii, T.: Improvement of interfacial adhesion in bamboo polymer composite enhanced with microfibrillated cellulose. Polym. Compos. Biocomposites **3**(1), 317–329 (2013). https://doi.org/10.1002/9783527674220.ch9
40. International Network for Bamboo & Rattan: Bamboo: a strategic resource for countries to reduce the effects of climate change. Policy Synthesis Report, 1–28 (2014)
41. Suhaily, S.S., Khalil, H.P.S.A., Nadirah, W.O.W., Jawaid, M.: Bamboo based biocomposites material, design and applications. Mater. Sci. **12**(21), 549–558 (2013). https://doi.org/10.5772/56700

42. Kavitha, S., Felix kala, T.: Study on structure and extraction of bamboo fiber. Asian J. Sci. Technol. **7**(2), 2426–2428 (2016)
43. Pulavarty, A., Sarangi, A.: Salt tolerance screening of bamboo genotypes (bamboo sps.) using growth and organic osmolytes accumulation as effective indicators. World Bamboo Congr **10**(1), 1–16 (2015)
44. Correal, J.F.: Bamboo design and construction. In: Nonconventional and Vernacular Construction Materials, 14(7), pp. 393–431 (2016). https://doi.org/10.1016/B978-0-08-100038-0.00014-7
45. Kim, H., Okubo, K., Fujii, T., Takemura, K.: Influence of fiber extraction and surface modification on mechanical properties of green composites with bamboo fiber. J. Adhes. Sci. Technol. **27**(12), 1348–1358 (2013). https://doi.org/10.1080/01694243.2012.697363
46. Yueping, W., Ge, W., Haitao, C., Genlin, T., Zheng, L., Feng, X.Q., Xiangqi, Z., Xiaojun, H., Xushan, G.: Structures of bamboo fiber for textiles. Text. Res. J. **80**(4), 334–343 (2010). https://doi.org/10.1177/0040517509337633
47. Bar-Yosef, O., Eren, M.I., Yuan, J., Cohen, D.J., Li, Y.: Were bamboo tools made in prehistoric Southeast Asia? An experimental view from South China. Quatern. Int. **26**(9), 9–21 (2012). https://doi.org/10.1016/j.quaint.2011.03.026
48. Khan, Z., Yousif, B.F., Islam, M.: Fracture behaviour of bamboo fiber reinforced epoxy composites. Compos. B Eng. **116**(12), 186–199 (2017). https://doi.org/10.1016/j.compositesb.2017.02.015
49. Zakikhani, P, Zahari, R., Sultan, M.T.H., Majid, D.L. Bamboo fibre extraction and its reinforced polymer composite material. Int. J. Chem. Biomol. Metall. Mater. Sci. Eng. **8**(4), 271–274 (2014). https://doi.org/10.14456/sjst.2014.16
50. Xiao, Y.: Engineered bamboo. In: Nonconventional and Vernacular Construction Materials, 15(9), pp. 433–452 (2016). https://doi.org/10.1016/B978-0-08-100038-0.00015-9
51. Yu, W.K., Chung, K.F., Chan, S.L.: Axial buckling of bamboo columns in bamboo scaffolds. Eng. Struct. **27**(1), 61–73 (2005). https://doi.org/10.1016/j.engstruct.2004.08.011
52. Bai, Y.Y., Xiao, L.P., Shi, Z.J., Sun, R.C.: Structural variation of bamboo lignin before and after ethanol organosolv pretreatment. Int. J. Mol. Sci. **14**(11), 21394–21413 (2013). https://doi.org/10.3390/ijms141121394
53. Resistance, C., Properties, T., Bamboo, O.F., Fibers, G., Epoxy, R., Composites, H.: Chemical resistance and tensile properties of bamboo and glass fibers reinforced epoxy hybrid composites. Int. J. Mater. Biomater. Appl. **1**(1), 17–20 (2011)
54. Khalil, H.A., Bhat, I.U.H., Jawaid, M., Zaidon, A., Hermawan, D., Hadi, Y.S.: Bamboo fibre reinforced biocomposites: a review. Mater. Des. **42**(3), 353–368 (2012). https://doi.org/10.1016/j.matdes.2012.06.015
55. Li, D.L., Wu, J.Q., Peng, W.X., Xiao, W.F., Wu, J.G., Zhuo, J.Y., Yuan, T.Q., Sun, R.C.: Effect of lignin on bamboo biomass self-bonding during hot-pressing: lignin structure and characterization. BioResources **10**(4), 6769–6782 (2015). https://doi.org/10.15376/biores.10.4.6769-6782
56. Zhang, Z., Xue, Q., Huang, K., Ma, Q., Guo, Y.: Study on dissociation of nano bamboo extractives. Extraction **4**(9), 4–7 (2013)
57. Xie, J., Hse, C.Y., Shupe, T.F., Pan, H., Hu, T.: Extraction and characterization of holocellulose fibers by microwave-assisted selective liquefaction of bamboo. J. Appl. Polym. Sci. **133**(18), 43–54 (2016). https://doi.org/10.1002/app.43394
58. Fiore, V., Di Bella, G., Valenza, A.: The effect of alkaline treatment on mechanical properties of kenaf fibers and their epoxy composites. Compos. B Eng. **68**(21), 14–21 (2015). https://doi.org/10.1016/j.compositesb.2014.08.025
59. Biswas, S., Ahsan, Q., Cenna, A., Hasan, M., Hassan, A.: Physical and mechanical properties of jute, bamboo and coir natural fiber. Fibers Polym. **14**(10), 1762–1767 (2013). https://doi.org/10.1007/s12221-013-1762-3
60. Nayak, L., Mishra, S.P.: Prospect of bamboo as a renewable textile fiber, historical overview, labeling, controversies and regulation. Fashion and Textiles **3**(1), 54–61 (2016). https://doi.org/10.1186/s40691-015-0054-5

61. Osorio, L., Trujillo, E., Van Vuure, A.W., Verpoest, I.: Morphological aspects and mechanical properties of single bamboo fibers and flexural characterization of bamboo/ epoxy composites. J. Reinf. Plast. Compos. **30**(5), 396–408 (2011). https://doi.org/10.1177/0731684410397683
62. Sugesty, S., Kardiansyah, T., Hardiani, H.: Bamboo as raw materials for dissolving pulp with environmental friendly technology for rayon fiber. Procedia Chem. **17**(23), 194–199 (2015). https://doi.org/10.1016/j.proche.2015.12.122
63. Pinho, E., Henriques, M., Oliveira, R., Dias, A., Soares, G.: Development of biofunctional textiles by the application of resveratrol to cotton, bamboo, and silk. Fibers Polym. **11**(2), 271–276 (2010). https://doi.org/10.1007/s12221-010-0271-x
64. Stelte, W.: Steam explosion for biomass pre-treatment. Danish Technological Institute (2013)
65. Yao, J., Bastiaansen, C., Peijs, T.: High strength and high modulus electrospun nanofibers. Fibers **2**(2), 158–186 (2014). https://doi.org/10.3390/fib2020158
66. Jayaramudu, J., Reddy, G.S.M., Varaprasad, K., Sadiku, E.R., Ray, S.S., Rajulu, A.V.: Mechanical properties of uniaxial natural fabric Grewia tilifolia reinforced epoxy-based composites: effects of chemical treatment. Fibers and Polym. **15**(7), 1462–1468 (2014). https://doi.org/10.1007/s12221-014-1462-7
67. Kang, J.T., Kim, S.H.: Improvement in the mechanical properties of polylactide and bamboo fiber biocomposites by fiber surface modification. Macromol. Res. **19**(8), 789–796 (2011). https://doi.org/10.1007/s13233-011-0807-y
68. Zhou, A., Huang, D., Li, H., Su, Y.: Hybrid approach to determine the mechanical parameters of fibers and matrixes of bamboo. Constr. Build. Mater. **35**(11), 191–196 (2012). https://doi.org/10.1016/j.conbuildmat.2012.03.011
69. Kuromi, Y., Sato, T., Ando, H., Matsumoto, Y., Oda, K., Ito, E., Ichikawa, M., Watanabe, T., Sakuma, J., Saito, K.: Removal of bamboo fragments transorbitally penetrated into the cerebellum and temporal lobe 30 years after the injury. Neurol. Surg. **40**(11), 979–983 (2012)
70. Kaur, V., Chattopadhyay, D.P., Kaur, S.: Study on extraction of bamboo fibres from raw bamboo fibres bundles using different retting techniques. Textiles and Industrial Science and Technology (TLIST) (2013)
71. Lin, J.S., Wang, X., Lu, G.: Crushing characteristics of fiber reinforced conical tubes with foam-filler. Compos. Struct. **116**(1), 18–28 (2014). https://doi.org/10.1016/j.compstruct.2014.04.023
72. Yu, Y., Huang, X., Yu, W.: A novel process to improve yield and mechanical performance of bamboo fiber reinforced composite via mechanical treatments. Compos. B Eng. **56**(8), 48–53 (2014). https://doi.org/10.1016/j.compositeb.2013.08.007
73. Correia, V. da C., dos Santos, V., Sain, M., Santos, S.F., Leão, A.L., Savastano Jr., H.: Grinding process for the production of nanofibrillated cellulose based on unbleached and bleached bamboo organosolv pulp. Cellul. **23**(5), 2971–2987 (2016). https://doi.org/10.1007/s10570-016-0996-9
74. Xie, J., Lin, Y.S., Shi, X.J., Zhu, X.Y., Su, W.K., Wang, P.: Mechanochemical-assisted extraction of flavonoids from bamboo (Phyllostachys edulis) leaves. Ind. Crops Prod. **43**, 276–282 (2013). https://doi.org/10.1016/j.indcrop.2012.07.041
75. Eriksson, M., Goossens, H., Peijs, T.: Influence of drying procedure on glass transition temperature of PMMA based nanocomposites. Nanocomposites **1**(1), 36–45 (2015). https://doi.org/10.1179/2055033214Y.0000000005
76. Li, Q., WenJi, Y., YangLun, Y.: Research on properties of reconstituted bamboo lumber made by thermo-treated bamboo bundle curtains. For. Prod. J. **62**(7/8), 545–550 (2012)
77. Peng, P., She, D.: Isolation, structural characterization, and potential applications of hemicelluloses from bamboo: a review. Carbohydr. Polym. **112**(67), 701–720 (2014). https://doi.org/10.1016/j.carbpol.2014.06.068
78. Meng, C., Yu, C.: Study on the oxidation degumming of ramie fiber. Chem. Mater. Metall. Eng. III, PTS, **1**(3), 881–883, 1497–1500 (2014). https://doi.org/10.4028/www.scientific.net/AMR.881-883.1497
79. Jaszkiewicz, A., Meljon, A., Bledzki, A.K.: Mechanical and thermomechanical properties of PLA/Man-made cellulose green composites modified with functional chain extenders: a comprehensive study. Polym. Compos. **8**(12), 41–56 (2018). https://doi.org/10.1002/pc.24122

80. Manalo, A.C., Wani, E., Zukarnain, N.A., Karunasena, W., Lau, K.T.: Effects of alkali treatment and elevated temperature on the mechanical properties of bamboo fibre-polyester composites. Compos. B Eng. **80**(3), 73–83 (2015). https://doi.org/10.1016/j.compositesb.2015.05.033
81. Jonoobi, M., Oladi, R., Davoudpour, Y., Oksman, K., Dufresne, A., Hamzeh, Y., Davoodi, R.: Different preparation methods and properties of nanostructured cellulose from various natural resources and residues: a review. Cellul. **22**(2), 935–969 (2015). https://doi.org/10.1007/s10570-015-0551-0
82. Trujillo, D., López, L.F.: Bamboo material characterisation. In: Nonconventional and Vernacular Construction Materials, 13(5), 365-392 (2016). https://doi.org/10.1016/B978-0-08-100038-0.00013-5.
83. Zakikhani, P., Zahari, R., Sultan, M.T.H., Majid, D.L.: Extraction and preparation of bamboo fibre-reinforced composites. Mater. Des. **63**(14), 820–828 (2014). https://doi.org/10.1016/j.matdes.2014.06.058
84. Castanet, E., Li, Q., Dumée, L.F., Garvey, C., Rajkhowa, R., Zhang, J., Rolfe, B., Magniez, K.: Structure–property relationships of elementary bamboo fibers. Cellul. **23**(6), 3521–3534 (2016). https://doi.org/10.1007/s10570-016-1078-8
85. Okubo, K., Fujii, T., Yamamoto, Y.: Development of bamboo-based polymer composites and their mechanical properties. Compos. A Appl. Sci. Manuf. **35**(3), 377–383 (2004). https://doi.org/10.1016/j.compositesa.2003.09.017
86. Phong, N.T., Fujii, T., Chuong, B., Okubo, K.: Study on how to effectively extract bamboo fibers from raw bamboo and wastewater treatment. J. Mater. Sci. Res. **1**(1), 144–152 (2011). https://doi.org/10.5539/jmsr.v1n1p144
87. Cromer, B.M., Coughlin, E.B., Lesser, A.J.: Evaluation of a new processing method for improved nanocomposite dispersions. Nanocomposites **1**(3), 152–159 (2015). https://doi.org/10.1179/2055033215Y.0000000009
88. Junior, M.G., Teixeira, F.G., Tonoli, G.H.D.: Effect of the nano-fibrillation of bamboo pulp on the thermal, structural, mechanical and physical properties of nanocomposites based on starch/poly (vinyl alcohol) blend. Cellul. **25**(3), 1823–1849 (2018). https://doi.org/10.1007/s10570-018-1691-9
89. Ghazy, A., Bassuoni, M., Maguire, E., O'Loan, M.: Properties of fiber-reinforced mortars incorporating nano-silica. Fibers **4**(1), 6–17 (2016). https://doi.org/10.3390/fib4010006
90. Jawaid, M.H.P.S., Khalil, H.A.: Cellulosic/synthetic fibre reinforced polymer hybrid composites: a review. Carbohyd. Polym. **86**(1), 1–18 (2011). https://doi.org/10.1016/j.carbpol.2011.04.043
91. Huda, S., Reddy, N., Yang, Y.: Ultra-light-weight composites from bamboo strips and polypropylene web with exceptional flexural properties. Compos. B Eng. **43**(3), 1658–1664 (2012). https://doi.org/10.1016/j.compositesb.2012.01.017
92. Liew, F.K., Hamdan, S., Rahman, M.R., Mahmood, M.R., Lai, J.C.H.: The effects of nanoclay and tin(IV) oxide nanopowder on morphological, thermo-mechanical properties of hexamethylene diisocyanate treated jute/bamboo/polyethylene hybrid composites. J. Vinyl Add. Tech. **67**(21), 789–794 (2017). https://doi.org/10.1002/vnl.21600
93. Rostamiyan, Y., Fereidoon, A., Mashhadzadeh, A.H., Ashtiyani, M.R., Salmankhani, A.: Using response surface methodology for modeling and optimizing tensile and impact strength properties of fiber orientated quaternary hybrid nano composite. Compos. B Eng. **69**(13), 304–316 (2015). https://doi.org/10.1016/j.compositesb.2014.09.031
94. Buonamici, F., Volpe, Y., Furferi, R., Carfagni, M., Signorini, G., Goli, G., Governi, L., Fioravanti, M.: Bamboo's bio-inspired material design through additive manufacturing technologies. Lect. Notes Civ. Eng. (2019). https://doi.org/10.1007/978-3-030-03676-8_32
95. Chen, H., Yu, Y., Zhong, T., Wu, Y., Li, Y., Wu, Z., Fei, B.: Effect of alkali treatment on microstructure and mechanical properties of individual bamboo fibers. Cellul. **24**(1), 11–16 (2017). https://doi.org/10.1007/s10570-016-1116-6
96. Reddy, K.R.: Polypropylene clay nanocomposites. In: Handbook of Polymer nanocomposites. Processing, Performance and Application: Volume A: Layered Silicates, 153–175 (2014). https://doi.org/10.1007/978-3-642-38649-7_2

97. Essabir, H., Boujmal, R., Bensalah, M.O., Rodrigue, D., Bouhfid, R., Qaiss, A.E.K.: Mechanical and thermal properties of hybrid composites: oil-palm fiber/clay reinforced high density polyethylene. Mech. Mater. **4**(2), 456–467 (2016). https://doi.org/10.1016/j.mechmat.2016.04.008
98. Silva, C.R., Lago, R.M., Veloso, H.S., Patricio, P.S.O.: Use of amphiphilic composites based on clay/carbon nanofibers as fillers in UHMWPE. J. Braz. Chem. Soc. **29**(2), 278–284 (2018). https://doi.org/10.21577/0103-5053.20170138
99. Kaur, N., Kishore, D.: Montmorillonite: an efficient, heterogeneous and green catalyst for organic synthesis. J. Chem. Pharm. Res. **4**(2), 991–1015 (2012)
100. Asad, M., Saba, N., Asiri, A.M., Jawaid, M., Indarti, E., Wanrosli, W.D.: Preparation and characterization of nanocomposite films from oil palm pulp nanocellulose/poly (Vinyl alcohol) by casting method. Carbohyd. Polym. **3**(15), 123–136 (2018). https://doi.org/10.1016/j.carbpol.2018.03.015
101. Voronova, M.I., Surov, O.V., Guseinov, S.S., Barannikov, V.P., Zakharov, A.G.: Thermal stability of polyvinyl alcohol/nanocrystalline cellulose composites. Carbohyd. Polym. **34**(56), 4590–4598 (2015). https://doi.org/10.1016/j.carbpol.2015.05.032
102. Tan, B.K., Ching, Y.C., Poh, S.C., Abdullah, L.C., Gan, S.N.: A review of natural fiber reinforced poly(vinyl alcohol) based composites: application and opportunity. Polym. **7**(11), 156–167 (2015). https://doi.org/10.3390/polym7111509
103. Mousa, M., Dong, Y., Davies, I.J.: Eco-friendly polyvinyl alcohol (PVA)/bamboo charcoal (BC) nanocomposites with superior mechanical and thermal properties. Adv. Compos. Mater **24**(30), 140–156 (2018). https://doi.org/10.1080/09243046.2017.1407906
104. Tang, C.M., Tian, Y.H., Hsu, S.H.: Poly(vinyl alcohol) nanocomposites reinforced with bamboo charcoal nanoparticles: mineralization behavior and characterization. Mater. **3**(9), 84–89 (2015). https://doi.org/10.3390/ma8084895
105. Zhang, X., Wang, F., Keer, L.M.: Influence of surface modification on the microstructure and thermo-mechanical properties of bamboo fibers. Mater. **8**(10), 6597–6608 (2015). https://doi.org/10.3390/ma8105327
106. Orue, A., Jauregi, A., Unsuain, U., Labidi, J., Eceiza, A., Arbelaiz, A.: The effect of alkaline and silane treatments on mechanical properties and breakage of sisal fibers and poly(lactic acid)/sisal fiber composites. Compos. A Appl. Sci. Manuf. **4**(1), 2121–2134 (2016). https://doi.org/10.1016/j.compositesa.2016.01.021
107. Oushabi, A., Sair, S., Oudrhiri Hassani, F., Abboud, Y., Tanane, O., El Bouari, A.: The effect of alkali treatment on mechanical, morphological and thermal properties of date palm fibers (DPFs): study of the interface of DPF–Polyurethane composite. S. Afr. J. Chem. Eng. **15**(5), 123–136 (2017). https://doi.org/10.1016/j.sajce.2017.04.005
108. Islam, M.S., Hamdan, S., Jusoh, I., Rahman, M.R., Ahmed, A.S.: The effect of alkali pretreatment on mechanical and morphological properties of tropical wood polymer composites. Mater. Des. **33**(2), 419–424 (2012). https://doi.org/10.1016/j.matdes.2011.04.044
109. Daneshpayeh, S., Ashenai Ghasemi, F., Ghasemi, I., Ayaz, M.: Predicting of mechanical properties of PP/LLDPE/TiO nano-composites by response surface methodology. Compos. B Eng. **84**(21), 109–120 (2016). https://doi.org/10.1016/j.compositesb.2015.08.075
110. Olivera, S., Muralidhara, H.B., Venkatesh, K., Gopalakrishna, K., Vivek, C.S.: Plating on acrylonitrile-butadiene-styrene (ABS) plastic: a review. J. Mater. Sci. **10**(8), 85–96 (2016). https://doi.org/10.1007/s10853-015-9668-7
111. Zhu, Y., Romain, C., Williams, C.K.: Sustainable polymers from renewable resources. Nat. **1**(2), 10–21 (2016). https://doi.org/10.1038/nature21001

Impact of Poly (Ethylene-Alt-Maleic Anhydride) and Nanoclay on the Physicochemical, Mechanical, and Thermal Properties of Bamboo Nanocomposite

Md Rezaur Rahman, Muhammad Adamu, Muhammad Khusairy Bin Bakri, and Siti Noor Linda bt Taib

Abstract The effect of poly (ethylene-alt-maleic anhydride) with nanoclay on bamboo nanocomposite was reported in this chapter. The Fourier transform infrared spectroscopy (FTIR), x-ray diffraction (XRD), scanning electron microscopy (SEM), differential scanning calorimetry (DSC) and thermogravimetric analysis (TGA) were used to characterize the nanocomposites. The mechanical test was conducted and reported in this chapter. The poly (ethylene-alt-maleic anhydride) with nanoclay enhanced the physical, mechanical and thermal properties of manufactured nanocomposites.

Keywords Poly (Ethylene-Alt-Maleic Anhydride) · Nanoclay · Nanocomposites · Bamboo · Properties

1 Introduction

For upgrading both of the synthetic polymers structural and functional properties, a new area of research among material scientists and engineer involving nanotechnology application has emerged [1, 2]. With the current challenge, there is a need to mitigate materials that are durable, sustainable, cost-effective, and environmentally friendly, even though the nanotechnology for production of new composite materials are well-recognized, especially on its synthesis and application [3]. Thus, there is considerable interest for the continuous search for low-cost reinforced composites using only biodegradables [4–7].

There is a proportional increase demand for wood, as the global economy grows in the world's [8]. Present data shows that sia-Pacific countries accounted approximately 24% of the global market, which indicate that wood trade has exceeded 1.8

M. R. Rahman (✉) · M. Adamu · M. K. B. Bakri · S. N. L. Taib
Faculty of Engineering, Universiti Malaysia Sarawak, Jalan Datuk Mohammad Musa, 94300, Kota Samarahan, Sarawak, Malaysia
e-mail: rmrezaur@unimas.my

M. R. Rahman (ed.), *Bamboo Polymer Nanocomposites*, Engineering Materials,
https://doi.org/10.1007/978-3-030-68090-9_2

billion m^3 [9]. The demand for high-quality hard wood has led to removal of non-renewable wood in many developing countries in Asia and has become a serious concern throughout the nations [10]. This development has led to a sharp rise in the cost of natural wood products, as many countries have taken measures to ban commercial logging.

Recently, the choice of natural fibers from timber is explored to replace some of these materials [11–13]. However, the mechanical properties are every so often not satisfied. Generally, in the advent of findings that are a combination of the matrix and natural fibers would yield composites with high strength-to-weight ratios, bamboo fibers have been a preferred due to their potential for the manufacture of materials that can be recyclable, biodegradable, and highly sustainable [14, 15]. Additionally, for optimum properties, each component of the matrix can effectively be utilized as required in the formation of a composite [16, 17]. To eases consumer choice and desirability, it has been established that high-end quality and sustainable industrial products can be formed from bamboo [18, 19]. In applications requiring tensile loading, bamboo fibers are often referred to as natural glass fiber, as a result of high strength-to-weight ratio, it is arguably an attractive substitute to steel [20–22].

Recently, due to its applications in industrial and scientific research, the synthesis of polymer-layered silicate (PLS) nanocomposite has been of great importance to produce value-added materials with highly improved physical and thermal properties [23, 24]. With improved physical and mechanical properties, synthesized of PLS nanocomposites have shown remarkable improvement, as compared with conventional micro or macro composites. For numerous engineering applications ranging from construction to household product, these properties required and include heat resistance, increased strength, rapid biodegradability, reduced gas absorptivity and flammability [25–27]. Compared to those of the original bulk polymer and also due to the dispersion of the polymer matrix and bamboo material as a nanoscale inorganic filler, these properties are re-engineered, consequently leading to an active interfacial area that translates to superior properties [28, 29]. Scientifically, the inorganic filler particles reduced to nanoscale dimensions, which may directly impact their properties [30].

Generally, the nature and compositions of the inorganic filler depend on the final properties of the nanocomposites formed, as well as the microstructure and interfacial interactions of the bamboo microstructure [31, 32]. Therefore, to produce a good nanocomposite material, the optimal inorganic filler that effectively interact and change the microstructure is remarkably essential step in the process [33, 34]. Compared to conventional filler polymers, talc, glass fibers, carbon black, and calcium carbonate particle is one of many clay nanofiller, which are usually referred to as micro size fillers, reported to have remarkable physical and mechanical properties [35]. For example, when clay nanofibers are used, the processability, mechanical properties, and lightness of the bamboo fiber can still be maintained and this has made it a preferred choice in the polymer industry [17, 36].

Clay nanofiller that is characteristic of silicate minerals possesses a well-layered structure [37, 38]. Due to its abundance, montmorillonite is the most common nanofiller among the various types of clay that have been used [39]. It contains alkali metal cations with a high surface area, great swelling capacity, strong cation exchange, and excellent absorptive properties [39]. Many studies have been made on composite and polymer blends, such as propylene ethylene (PE) [40]. However, because of their difference in polarity, most of these plastics become immiscible. Numerous efforts have been deployed to mitigate this compatibility issue, especially to modify either the polymer or the composite [41]. However, there is some compatibilizers, which have been used resulted in weak and unsatisfactory mechanical properties [41]. Weak interaction between the compatibilizers and the composite attributed to the lower mechanical properties [42]. Recently, polymers containing reactive groups such as maleic anhydride has gained attention [41, 43]. In the composite, the anhydride groups react with the hydroxyl groups present in the composite by chemical bonding, which create dispersion of the composite, enhanced interfacial adhesion, and subsequently improved mechanical properties.

There are several chemical and surface modification techniques, which include impregnation, salinization, acetylation, benzoylation, corona/cold plasma maleization, peroxide, enzymatic, or isocyanate [3]. Therefore, the main objective of this study is to develop a bamboo nanocomposite material. It also covered the investigation on the effect of poly (ethylene-alt-maleic anhydride) (PEA) as compatibilizer and nanoclay on the physical, thermal, morphological, and mechanical properties of the developed nanocomposite. Poly (ethylene-alt-maleic anhydride) and nanoclay were used to improve the physicochemical, thermal, and mechanical properties of bamboo nanocomposite by impregnation under vacuum technique. The raw bamboo (RB) and nanocomposites have been characterized using Fourier transform infrared (FTIR), x-ray diffraction (XRD), and scanning electron microscopy (SEM) analyses. The thermal and mechanical properties were also reported in this study.

2 Methodology

2.1 Materials

The bamboo (*gigantochloa scortechenii*) was obtained from a forest in Kota Samarahan, Sarawak, Malaysia. Analytical grade chemical of ethanol and poly (ethylene-alt-maleic anhydride) (PEA) (99% purity, Sigma-Aldrich, St. Louis, MO, USA), sodium hydroxide and benzoyl peroxide (Merck Schuchardt OHG, Germany), and nanoclay (Cloisite Na^+, BYK, Wesel, Germany) were also used. Ethanol was used as a solvent to dissolve the radicals, sodium hydroxide was used to adjust the pH of the medium for enhanced polymerization, and benzoyl peroxide was used as catalyst.

2.2 Samples Preparation

To prepare the specimen, bamboo was cut into strips pieces and cleaned impregnation. The bamboo strips were placed in a forced air convection heater (Impact Test Equipment Ltd., Stevenston Ayrshire, Scotland) for five days at 70 °C for conditioning and drying. This reduced or eliminated the water present in the bamboo.

2.3 Samples Impregnation

The oven-dried bamboo strips were immersed in a solution. It was prepared by adding the different amounts of montmorillonite nanoclay (calosite), sodium hydroxide, polymers, and 10 mg of benzoyl peroxide dissolved in 500 mL of ethanol. The bamboo and the solution were then transferred into a vacuum chamber at different times of impregnation. The impregnated specimens in the vacuum chamber were then later removed. In each impregnation, five replicate samples were produced.

2.4 Polymerization and Curing

The comopsites surfaces were cleaned with tissue paper, covered with aluminum foil, and placed into an oven operating at a temperature of 80 °C for 48 h. This is to allow polymerization and curing process of the bamboo fiber and cross-linkage of the nanoclay.

2.5 Mechanical Testing

2.5.1 Three-Point Bending Test

The samples dimensions are 30 mm (L) × 20 mm (T) × 20 mm (W) in accordance with ASTM D790-17 [44]. The modulus of rupture (MOR) and modulus of elasticity (MOE) for RB and TB samples were then calculated. Three-point bending tests were conducted using a Shimadzu MSC-5/500 universal testing machine (Kyoto, Japan) operating at a crosshead speed of 5 mm/min. The MOR and MOE were calculated using Eqs. (1) and (2):

$$MOR = 1.5LW/bd^2 \tag{1}$$

$$MOE = L^2m/4bd^3 \tag{2}$$

Where, W is the maximum load for wood failure, L is the distance between centers of support, b is the mean width (tangential direction) of the sample, d is the mean thickness (radial direction) of the sample, and m is the slope of the tangent to the initial line of the force–displacement curve.

2.6 *Characterization for Physico-Chemical and Morphology*

2.6.1 FTIR Spectroscopy

The infrared spectra of all samples were obtained using a Shimadzu IRAffinity^{-1} spectrophotometer (Shimadzu, Kyoto, Japan). A wavenumber ranges from 4000 to 400 cm^{-1} was used. This analytical tool is used excellently in investigating polymers, clays, and clay minerals to identify the molecular bond structures and the functional groups of the specimens [37]. IRsolution software was used to plot the FTIR spectrum based on ASTM E168-16 [45] and ASTM E1252-98 [46] standards.

2.6.2 X-ray Diffraction (XRD)

The structure and crystallinity of the RB and the prepared nanocomposites were characterized using XRD analysis. The pulverized specimen diffractograms were obtained using a Bruker D8 advanced x-ray diffractometer (Bruker Optik GmbH, Ettlingen, Germany) with CuKα radiation ($\lambda = 1.5418$ Å, rated as 1.6 kW). It has a diffract meter from of 2θ, from 0° to 90°. An experimental procedure for the determination of degree of crystallinity is acceptable to these materials in as much as X-ray scattering curves are resolvable into crystalline and amorphous scattering regions. The crystallinity index (CI_{XRD}) was calculated using Eq. (1) from the height ratio of diffraction peaks:

$$I_{XRD} = \frac{I_{002-I_{am}}}{I_{002}} * 100 \tag{3}$$

where, I_{002} is the highest intensity of the peak at 2θ at about 22° and I_{am} is the lowest intensity of baseline at 2θ at about 30°, which corresponds to crystalline and amorphous parts, respectively. The d-spacing between the [002] lattice planes (d_{002}) of the sample was calculated using the Bragg Eq. (4):

$$d_{002=\frac{n*\lambda}{2Sin\theta}} \tag{4}$$

2.6.3 Scanning Electron Microscopy (SEM)

The morphological images were taken using Hitachi Analytical tablet top SEM (benchtop) TM-3030 (Hitachi High-Technologies (Germany) Europe GmbH.). It was done by mounting the sliced samples on the aluminum stubs, which was later fine coated using 'JFC-1600' (JEOL (Japan) Ltd.). The fine coated spurring the particle metal coating (i.e. gold) for 1–2 min, which gives a thickness of 10 nm under 0.1 torr and 18 mA. The collected images of the surface of composites were taken using a field emission gun with accelerating voltage of 20 kV. The tests were conducted according to the ASTM E2015-04 [47] standard.

2.7 Thermal Testing

2.7.1 Differential Scanning Calorimetry (DSC)

DSC Q10 (TA instrument) thermal system with the use of a sealed aluminum capsule was employed for differential scanning calorimetry (DSC) measurements. Each data represented a mean of five repeated runs. The specimen was weighed to about 4–4.5 mg and held at a single heating rate of 10 ºC/min and scanning temperature from 30 to 450 °C [48]. The crystallization and melting temperatures of the samples were obtained. This is in accordance to ASTM D3418-15 [49] and ASTM E1269-11 [50] standards.

2.7.2 Thermogravimetric Analysis (TGA)

Thermogravimetric analysis (TGA) was conducted using a TA Instruments 9222 thermal analyzer (TA Instruments, New Castle, DE, USA) with a platinum sample pan. The analysis was completed under nitrogen gas. Composites of 10 mg were analyzed at a maintained heating rate of 20 °C/min and heated to 600 °C. The test was in accordance with ASTM E1131-20 [51] standard.

3 Results and Disucssions

3.1 FTIR Analysis

The FTIR result of the RB and nanocomposites shown in Fig. 1. The main functional groups, Al–OH and Si–O were observed in the range of 1000–500 cm^{-1}. While, the dominant peaks of OH-stretching and CH-stretching at were observed in the entire spectra approximately 3444 cm^{-1} and 2907 cm^{-1}, respectively. Respectively, these

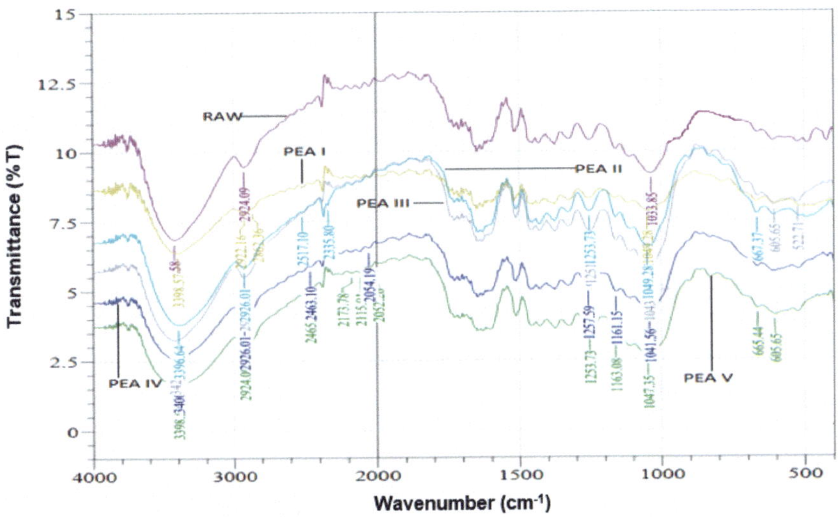

Fig. 1 FTIR spectra of RB and nanocomposites

peaks were also associated with the H–O–H stretching vibration of absorbed water in carbohydrates, the C–O stretching, and the C1-H deformation vibrations of cellulose. However, this peak completely disappeared in the spectra of nanocomposites, which could be attributed to the removal of hemicellulose after NaOH treatment. The peak intensities between 667 to 522.7 cm^{-1}, and 1092 to 1029 cm^{-1} resulted from the out-of-plane bending vibrations of –OH groups and stretching vibrations of Al–O–Si of the clay in the bamboo [52–55]. Therefore, for the nanocomposites, the strength of this band was higher than RB because of the bending vibrations of the AL–O–Si bond [56]. The peaks between 4000 and 3000 cm^{-1} showed broadening vibration of hydrogen bonds in –OH groups, whereas the stretching of C-H in methylene and methyl groups occurred between 3000 and 2900 cm^{-1} [56]. This change occurred is due to the improved number of double bonds of the aromatics present in PEA [57]. As it plasticized within the bamboo lumens, the nanoclay displaced the free-bound water content of the bamboo matrix [32]. The change in peak intensities from 3426 to 3396 cm^{-1}, 1091 to 1029 cm^{-1}, and 667 to 522.7 cm^{-1} showed that nanoclay was dispersed into the bamboo matrix [58]. The peaks at 1507 and 1378 cm^{-1} in the spectrum of the original particles were attributed to the C = C stretching vibration and C–H deformation vibration in the aromatic ring of lignin [59] disappeared completely on the remaining spectra, respectively. From the newly prepared nanocomposites, this indicates that lignin was removed. After the modification, the hemicellulose and lignin were largely removed and that the original molecular structure of the cellulose was maintained even after the removal of the matrix components.

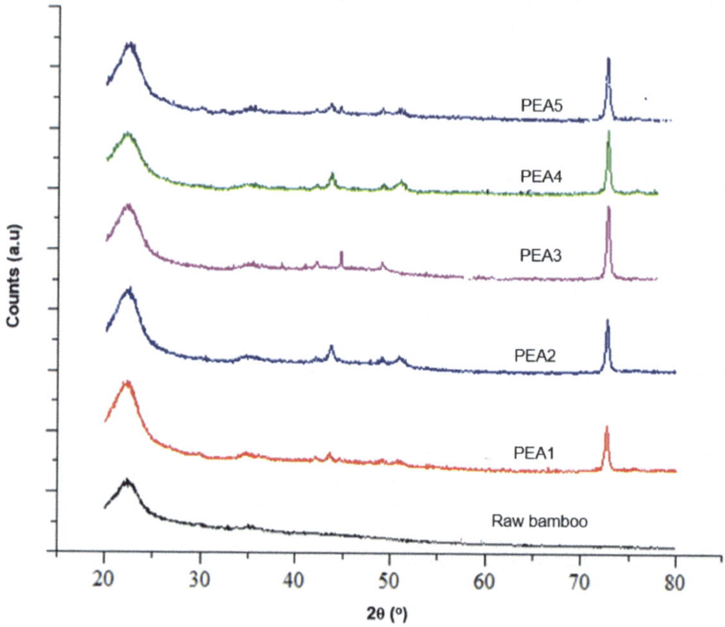

Fig. 2 XRD diffractograms of RB and PEA1, PEA2, PEA3, PEA4, and PEA5 nanocomposites

3.2 *XRD Analysis*

For identifying the structure of the RB and that of the nanocomposites, x-ray diffraction tool is used. The result is presented in Fig. 2. The nanocomposite with PEA showed peaks at 2θ of 23°, 43°, and 73° representing (001), (220), and (311) reflections, respectively [60]. Due to change in the structure of bamboo from an amorphous phase into a crystalline phase by the polymer matrix and nanoclay, the transformation has occurred. The raw bamboo exhibited a peak at only angle 23º, which is the characteristic peak of amorphous bamboo. The XRD pattern of the nanocomposite indicated the intercalation and dispersion of polymer and clay into the bamboo, which resulted in the pronounced peaks [55]. These increases in peak intensities could be related to the high amount of coupling agent that causes the good interaction between the polymer chains and nanoclay. In other words, the good dispersion of nanoclay in the polymer matrix presence as the compatibilizer.

3.3 *Morphological Properties*

The SEM micrographs in Fig. 3 showed the adhesion of polymer-filled lumens via *in-situ* polymerization and dispersion of nanoclay into the voids of the bamboo. For

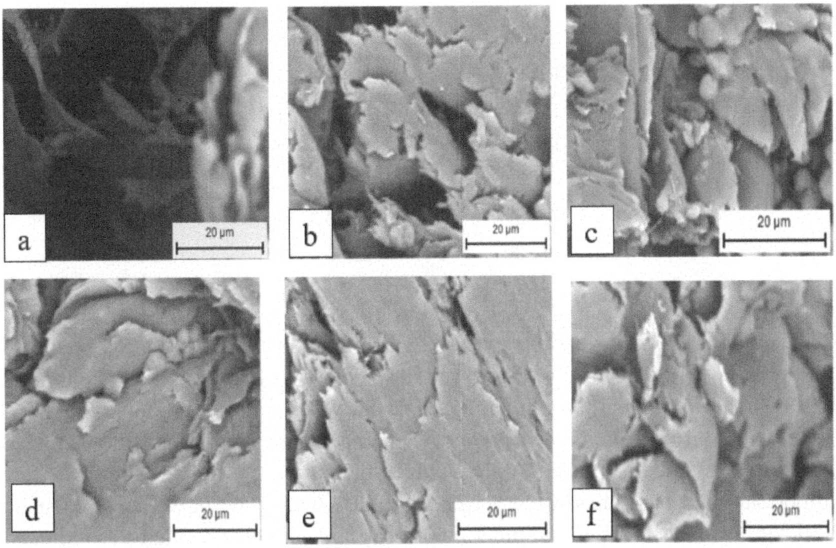

Fig. 3 SEM images (1500×) of: **a** raw bamboo, **b** PEA1, **c** PEA2, **d** PEA3, **e** PEA4, and **f** PEA5

the nanocomposites in Fig. 3b through 3(f), the surfaces observed were smoother than the RB surface in Fig. 3a, which had visible pore spaces. This was due to the polymer and nanoclay interaction that acted as filler materials, which filled the void spaces between the polymer and bamboo fiber pores [61]. When compared to the RB, the covalent bonds between the PEA and bamboo fiber walls were formed, which in turn filled the nanocomposites' void spaces. As a result of this interaction, a more rigid composite was formed, which showed compatibility and translated into improved mechanical properties of the composite [17, 62]. The compaction in the nanocomposites indicated that water could not easily flow into the bamboo. Thus, the treatment addressed hydrophobicity associated with bamboo and therefore extended its durability and external usage. The changes in morphology of both the RB and nanocomposites are illustrated in the SEM images in Fig. 3. Depending on the nanocomposite's composition, domains of a high homogeneity of clay and PEA distribution, no agglomerates were found. In all the cases a homogeneous distribution of the nanoclay in the PEA matrix, as produced by, may be seen.

3.4 Mechanical Properties

Table 1 shows that the values of MOE and MOR of the nanocomposites showed significant improvement compared to RB, whereas the highest values obtained is the PEA3 treatment. Modification of the RB indicated an increase in the elastic property, especially when the bamboo was treated, in which it is equally pointed out that

Table 1 MOE and MOR of RB and nanocomposites

Samples	MOE (GPa)	MOR (MPa)
Raw bamboo	7.82 ± 2.04	68.67 ± 27.29
PEA1	14.97 ± 3.02	102.10 ± 7.32
PEA2	14.57 ± 3.52	90.85 ± 10.27
PEA3	18.96 ± 3.09	121.48 ± 8.54
PEA4	11.20 ± 3.47	87.39 ± 22.15
PEA5	12.66 ± 2.20	12.54

it could withstand great loading pressure when used in construction applications. Additionally, the treated bamboo showed an increase of approximately 77% in its MOR, which was higher than similar materials like wood and timber sources obtained by Sultan et al. [63]. Therefore, to improve its mechanical properties, the nanoclay and PEA used were dispersed into the voids of the bamboo. Compared with the raw bamboo samples, bamboo samples treated with PEA/nanoclay showed higher tensile and flexural values. Further combination of PEA and nanoclay formed a crosslinked structure with bamboo cell wall through its hydroxyl groups, while PEA enhanced mechanical properties significantly [64]. The surface hydroxyl groups present interacted with hydroxyl of the bamboo fiber in modified nanocomposites, whereas the crosslinker and prepolymer resulting in enhanced properties. The nanoclay layers restricted the mobility of the polymer chains as they were fastened in between its gallery layers. Hence stiffened the composites, which improve the MOE and MOR.

4 Thermal Properties

4.1 DSC Analysis

Figure 4 shows the DSC curves of the RB and formed nanocomposites. Figure 4 indicated that RB exhibited three separated endothermic peaks at 80 °C to 110 °C, 180 °C to 220 °C, and 330 °C to 400 °C, which corresponded to amorphous, para-crystalline, and crystalline phases, respectively [65, 66]. Compared to those of the crystalline components, the para-crystalline parts and the amorphous sections were more susceptible to heat and chemicals. The non-crystalline molecular movement phase was higher compared to the crystalline part [48, 67]. While, the chain movement of macromolecules was firm. This change in the movement was ascribed to the intramolecular forces present as well as intermolecular hydrogen bonding [68].

The first endothermic peak is corresponded to the removal of absorbed water from the amorphous RB and the nanocomposites. Compared to RB, the area of the peak of the nanocomposites was smaller, which was due to the filler in the nanocomposites that inhibited the amorphous change, whereas reduced the amount of water absorbed into the RB. Because of the ionic characteristics of nanoclay,

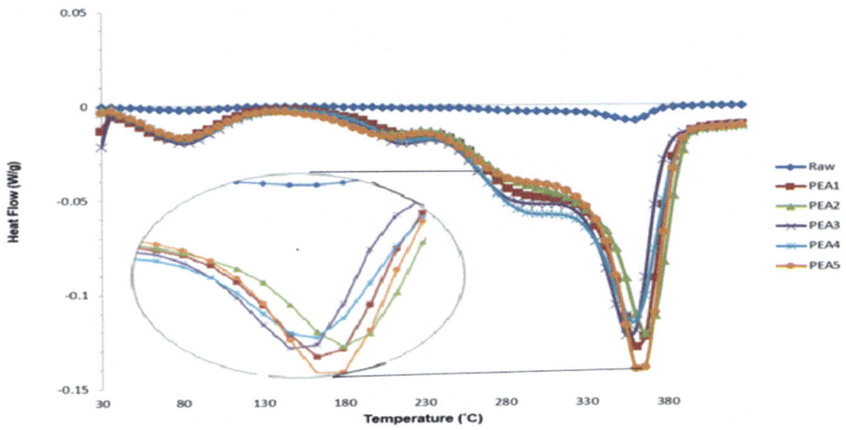

Fig. 4 DSC thermogram of **a** RB, **b** PEA1, **c** PEA2, **d** PEA3, **e** PEA4, and **f** PEA5

the hydrophobicity of the nanocomposites would be higher than that of the RB, which made the water molecule to be tightly held together. The thermal stability of the second endothermic para-crystalline phase of the RB and nanocomposites was similar, whereas the para-crystalline enthalpy of RB was higher than that of the nanocomposites. The clay had no covalent bond with the bamboo but restricted the para-crystalline movement to increase the crystallinity of the composite. With the aromatic groups in the bamboo, the molecule formed an induced dipole bond. The bamboo amorphous structure was then converted to a para-crystalline structure. The para-crystalline endothermic enthalpy of the nanocomposites was lower than RB because PEA and clay entered the para-crystalline region of the bamboo to convert it into a crystalline region [69]. The nanoclay particle entered the amorphous region and restricted the amorphous movement of the molecule. The crystallinity of the nanocomposites was higher compared to the RB.

4.1.1 Thermogravimetric Analysis

Figure 5 revealed three stages of thermal decomposition and degradation of RB and its nanocomposites. Stage 1 showed that the weight loss of all samples due to the evaporation of the absorbed moisture, which occurred at 72 °C [48, 70]. Below 110 °C, the initial weight loss of the samples was lowest for PEA5, followed by PEA4, PEA2, PEA3, PEA1, and RB, which is due to the result of the hydrolyzed reaction with the –OH groups of bamboo wall, as well as the surface modified clay filler, which also filled the cavities of the formed nanocomposites.

The second stage of thermal decomposition occurred at 125 °C, with an appreciable difference between the RB and its nanocomposites at 250 °C, which is due to the strong covalent bonds in the nanocomposite samples. This made weight loss noticeably lower in the nanocomposite samples compared to the RB like the second

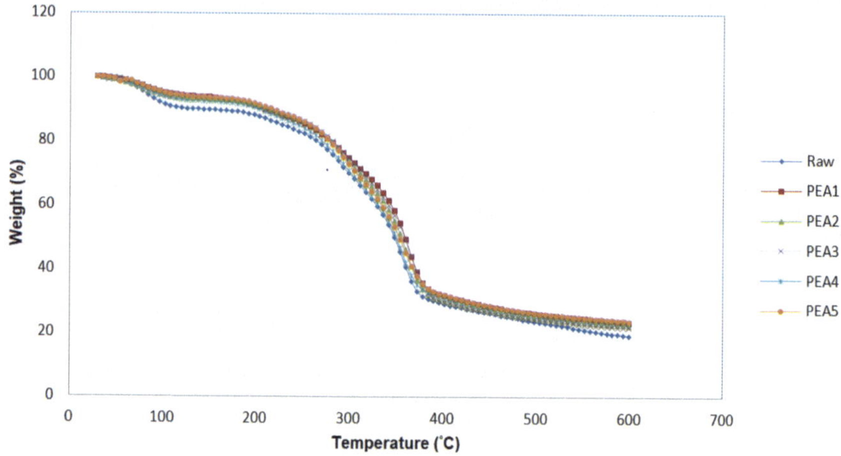

Fig. 5 TGA of **a** RB, **b** PEA1, **c** PEA2, **d** PEA3, **e** PEA4, and **f** PEA5

stage. Additionally, the third stage of thermal decomposition, which occurred at 385 °C revealed an appreciable difference between the RB and nanocomposite samples, whereas the decomposition being lower in the nanocomposites compared to the RB. This occurrence attributed to well-distributed clay layers, which blocking the way of unstable decomposition products, throughout the composites during thermal degradation. Thus, the remarkable improvement in thermal stability for nanocomposites formed during the third stage of decomposition [71, 72]. Thermal properties of all samples are presented in Table 2, whereas all the thermal degradation stages outlined the raw bamboo had the highest percentage of weight loss compared to the developed nanocomposites.

5 Conclusions

In a nutshells, bamboo nanocomposites were successfully prepared by using cloisite Na^+ nanoclay and PEA as a filler and compatibilizer. Compared to the raw bamboo, the nanocomposites demonstrated higher MOE and MOR. It was also observed incorporation of 15 g of nanoclay and 10 mg of PEA at a pH of 9, causes the modified bamboo had the highest increase of MOE and MOR. There was new modification of composition, which form from the raw bamboo resulted in improved mechanical properties of the prepared nanocomposites as shown by XRD and FTIR. The SEM study revealed the presence of nanoclay in the lumen, void spaces, and cell wall of the bamboo, improving adhesion between the polymer and the bamboo. Hence, it may reduce it hydrophobicity. The TGA and DSC analyses showed improved thermal stability of the nanocomposites compared to the raw bamboo. The improvement in

Table 2 Result of TGA for RB and nanocomposites

Sample	No. of transition	Transition temperature (°C)			Weight loss (%)
		T_i	T_m	T_f	
RAW	1	35.80	60.24	101.54	8.20
	2	117.21	146.35	169.70	44.71
	3	176.58	354.51	578.59	39.26
PEA1	1	41.42	78.36	127.26	5.62
	2	273.65	354.05	382.72	30.75
	3	390.14	429.87	508.14	28.94
PEA2	1	41.46	79.36	129.16	5.44
	2	274.24	351.20	381.98	21.02
	3	390.65	428.48	507.85	26.73
PEA3	1	42.01	77.35	127.02	5.57
	2	273.65	351.22	383.00	23.05
	3	389.76	429.09	506.92	28.08
PEA4	1	43.12	80.21	133.46	5.43
	2	273.11	352.42	382.07	30.69
	3	389.71	429.06	506.24	27.97
PEA5	1	41.39	80.02	128.22	5.24
	2	274.01	353.11	380.24	31.34
	3	390.13	427.58	503.49	28.48

T_i is initial temperature, T_m is temperature at which the rate of weight loss was maximum, and T_f is the final temperature

thermal properties of the modified bamboo was attributed to good dispersion of the nanoclay and PEA polymer matrix into the bamboo.

Acknowledgements The authors would like to acknowledge Universiti Malaysia Sarawak (UNIMAS) for the support.

References

1. Olad, A.: Polymer/clay nanocomposites. In: Reddy, B. (ed.), Advances in Diverse Industrial Applications of Nanocomposites, pp. 113–138. InTechOpen, London, UK (2011). https://doi.org/10.5772/14464
2. Shahadat, M., Teng, T.T., Rafatullah, M., Arshad, M.: Titanium-based nanocomposite materials: a review of recent advances and perspectives. Colloid. Surface. B **126**, 121–137 (2015). https://doi.org/10.1016/j.colsurfb.2014.11.049
3. Muhammad, A., Rahman, M.R., Hamdan, S., Sanaullah, K.: Recent developments in bamboo fiber-based composites: a review. Polym. Bull. **76**(5), 2655–2682 (2019). https://doi.org/10.1007/s00289-018-2493-9
4. Yates, M.R., Barlow, C.Y.: Life cycle assessments of biodegradable, commercial biopolymers—a critical review. Resour. Conserv. Recycl. **78**, 54–66 (2013). https://doi.org/10.1016/j.resconrec.2013.06.010

5. Thakur, V.K., Thakur, M.K., Gupta, R.K.: Review: raw natural fiber-based polymer composites. Int. J. Polym. Anal. Charact. **19**(3), 256–271 (2014). https://doi.org/10.1080/1023666X.2014.880016
6. Lu, N., Oza, S., Tajabadi, M.G.: Surface modification of natural fibers for reinforcement in polymeric composites. In: Thakur, V.K., Singha, A.S. (eds.) Surface Modification of Biopolymers, pp. 224–237. Wiley, Hoboken, NJ (2015). https://doi.org/10.1002/9781119044901.ch9
7. Varghese, A. M., Mittal, V.: Polymer composites with functionalized natural fibers. In: Shimpi, N.G. (ed.) Biodegradable and Biocompatible Polymer Composites, pp. 157–186. Woodhead Publishing, Cambridge, UK (2018). https://doi.org/10.1016/B978-0-08-100970-3.00006-7
8. Bais, A.L.S., Lauk, C., Kastner, T., Erb, K.: Global patterns and trends of wood harvest and use between 1990 and 2010. Ecol. Econ. **119**, 326–337 (2015). https://doi.org/10.1016/j.ecolecon.2015.09.011
9. Buongiorno, J., Raunikar, R., Zhu, S.: Consequences of increasing bioenergy demand on wood and forests: an application of the Global Forest Products Model. J. Forest Econ. **17**(2), 214–229 (2011). https://doi.org/10.1016/j.jfe.2011.02.008
10. Warman, R.D.: Global wood production from natural forests has peaked. Biodivers. Conserv. **23**(5), 1063–1078 (2014). https://doi.org/10.1007/s10531-014-0633-6
11. Saiful Islam, M., Hamdan, S., Jusoh, I., Rezaur Rahman, M., Ahmed, A.S.: The effect of alkali pretreatment on mechanical and morphological properties of tropical wood polymer composites. Mater. Des. **33**, 419–424 (2012). https://doi.org/10.1016/j.matdes.2011.04.044
12. Zhu, Y., Romain, C., Williams, C.K.: Sustainable polymers from renewable resources. Nature **540**, 354–362 (2016). https://doi.org/10.1038/nature21001
13. Hossen, M.F., Hamdan, S., Rahman, M.R.: Improved mechanical properties of silane treated jute/polyethylene/clay nanocomposites. Malays. Appl. Biol. **47**(1), 209–216 (2018)
14. Huda, S., Reddy, N., Yang, Y.: Ultra-light-weight composites from bamboo strips and polypropylene web with exceptional flexural properties. Compos. Part B-Eng. **43**(3), 1658–1664 (2012). https://doi.org/10.1016/j.compositesb.2012.01.017
15. Imbulana, P.K., Fernandez, T., Jayawardene, P.A.R.P., Levangama, T.P., Perera, Y.K., Arachchi, H.N.K., Mallawaarachchi, R.S.: Bamboo as a low cost and green alternative for reinforcement in light weight concrete. In: SAITM Research Symposium on Engineering Advancements 2013 (SAITM –RSEA 2013), pp. 166–172. Malabe, Sri Lanka (2013)
16. Loh, Y.R., Sujan, D., Rahman, M.E., Das, C.A.: Sugarcane bagasse—the future composite material: a literature review. Resour. Conserv. Recycl. **75**, 14–22 (2013). https://doi.org/10.1016/j.resconrec.2013.03.002
17. Gheith, M.H., Aziz, M.A., Ghori, W., Saba, N., Asim, M., Jawaid, M., Alothman, O.Y.: Flexural, thermal and dynamic mechanical properties of date palm fibres reinforced epoxy composites. J. Mater. Res. Technol. **8**(1), 853–860 (2019). https://doi.org/10.1016/j.jmrt.2018.06.013
18. Nahar, S., Hasan, M.: Effect of chemical composition, anatomy and cell wall structure on tensile properties of bamboo fiber. Eng. J. **17**(1), 61–68 (2013). https://doi.org/10.4186/ej.2013.17.1.61
19. Anokye, R., Bakar, E.S., Jegatheswaran, R., Awang, K.B.: Bamboo properties and suitability as a replacement for wood. Pertanika J. Sch. Res. Rev. **2**(1), 64–80 (2016). https://doi.org/10.13140/RG.2.1.1939.3048
20. Jawaid, M., Abdul Khalil, H.P.S.: Cellulosic/synthetic fibre reinforced polymer hybrid composites: a review. Carbohyd. Polym. **86**(1), 1–18 (2011). https://doi.org/10.1016/j.carbpol.2011.04.043
21. Karthik, S., Rao, P.R.M., Awoyera, P.O. Strength properties of bamboo and steel reinforced concrete containing manufactured sand and mineral admixtures. J. King Saud Univ.-Eng. Sci. **29**(4), 400–406 (2017). https://doi.org/10.1016/j.jksues.2016.12.003
22. Siddique, S.F., Priyanka, S., Nishanth, L.: Behavior of reinforced cement concrete beam with bamboo as partial replacement for reinforcement. Int. J. Civ. Eng. Technol. **8**(9), 580–587 (2017)
23. Ahmad, M., Kamke, F.A.: Properties of parallel strand lumber from Calcutta bamboo (Dendrocalamus strictus). Wood Sci. Technol. **45**(1), 63–72 (2011). https://doi.org/10.1007/s00226-010-0308-8

24. Shipp, D.A.: Polymer-layered silicate nanocomposites. In: Andrews, D.L., Scholes, G.D., Wiederrecht, G.P. (eds.) Comprehensive Nanoscience and Technology, pp. 265–276. Academic Press, Cambridge, MA (2011). https://doi.org/10.1016/B978-0-12-374396-1.00058-1
25. Biswas, M., Ray, S.S.: Recent progress in synthesis and evaluation of polymer-montmorillonite nanocomposites. In: New Polymerization Techniques and Synthetic Methodologies, pp. 167–221. Springer-Verlag, Berlin (2001). https://doi.org/10.1007/3-540-44473-4_3
26. Ray, S.S., Yamada, K., Okamoto, M., Ueda, K.: Polylactide-layered silicate nanocomposite: a novel biodegradable material. Nano Lett. **2**(10), 1093–1096 (2002). https://doi.org/10.1021/nl0202152
27. Thomas, S.,Kuruvilla, J., Malhotra, S.K., Goda, K., Sreekala, M.S.: Polymer Composites, vol. 1. Wiley-VCH, Weinheim (2012). https://doi.org/10.1002/9783527645213
28. Zhang, X., Wang, F., Keer, L.M.: Influence of surface modification on the microstructure and thermo-mechanical properties of bamboo fibers. Materials **8**(10), 6597–6608 (2015). https://doi.org/10.3390/ma8105327
29. Wang, G., Chen, F.: Development of bamboo fiber-based composites. In: Fan, M., Fu, F. (eds.) Advanced High Strength Natural Fibre Composites in Construction, pp. 235–255. Woodhead Publishing, Cambridge, UK (2017). https://doi.org/10.1016/B978-0-08-100411-1.00010-8
30. Sasthiryar, S., Abdul Khalil, H.P.S., Bhat, A.H., Ahmad, Z.A., Islam, M.N., Zaidon, A., Dungani, R.: Nanobioceramic composites: a study of mechanical, morphological, and thermal properties. BioResources **9**(1), 861–871 (2014). https://doi.org/10.15376/biores.9.1.861-871
31. Yu, Y.-L., Huang, X.-A., Yu, W.-J.: High performance of bamboo-based fiber composites from long bamboo fiber bundles and phenolic resins. J. Appl. Polym. Sci. **131**(12), Article ID 40371 (2014). https://doi.org/10.1002/app.40371
32. Liew, F.K., Hamdan, S., Rahman, M.R., Rusop, M.: Thermomechanical properties of jute/bamboo cellulose composite and its hybrid composites: the effects of treatment and fiber loading. Adv. Mater. Sci. Eng. Article ID 8630749 (2017). https://doi.org/10.1155/2017/8630749
33. Reddy, K.R.: Polypropylene clay nanocomposites. In: Pandey, J.K., Reddy, K.R., Mohanty, A.K., Misra, M. (eds.) Handbook of Polymernanocomposites. Processing, Performance and Application: Volume A: Layered Silicates, pp. 153–175, Springer-Verlag, Berlin and Heidelberg (2014). https://doi.org/10.1007/978-3-642-38649-7_2
34. Chen, H., Yu, Y., Zhong, T., Wu, Y., Li, Y., Wu, Z., Fei, B.: Effect of alkali treatment on microstructure and mechanical properties of individual bamboo fibers. Cellul. **24**(1), 333–347 (2017). https://doi.org/10.1007/s10570-016-1116-6
35. Thostenson, E.T., Li, C., Chou, T.-W.: Nanocomposites in context. Compos. Sci. Technol. **65**(3–4), 491–516 (2005). https://doi.org/10.1016/j.compscitech.2004.11.003
36. Meng, L.-Y., Park, S.-J.: Influence of carbon nanofibers on electrochemical properties of carbon nanofibers/glass fibers composites. Curr. Appl. Phys. **13**(4), 640–644 (2013). https://doi.org/10.1016/j.cap.2012.10.008
37. Jonoobi, M., Oladi, R., Davoudpour, Y., Oksman, K., Dufresne, A., Hamzeh, Y., Davoodi, R.: Different preparation methods and properties of nanostructured cellulose from various natural resources and residues: a review. Cellul. **22**(2), 935–969 (2015). https://doi.org/10.1007/s10570-015-0551-0
38. Silva, C.R., Lago, R.M., Veloso, H.S., Patricio, P.S.O.: Use of amphiphilic composites based on clay/carbon nanofibers as fillers in UHMWPE. J. Braz. Chem. Soc. **29**(2), 278–284 (2018). https://doi.org/10.21577/0103-5053.20170138
39. Kaur, N., Kishore, D.: Montmorillonite: an efficient, heterogeneous and green catalyst for organic synthesis. J. Chem. Pharm. Res. **4**(2), 991–1015 (2012)
40. Duy Tran, T., DangNguyen, M., Ha Thuc, C.N., Ha Thuc, H., Dang Tan, T.: Study of mechanical properties of composite material based on polypropylene and Vietnamese rice husk filler. J. Chem., Article ID 752924 (2013). https://doi.org/10.1155/2013/752924
41. Liu, W., Wang, Y.-J., Sun, Z.: Effects of polyethylene-grafted maleic anhydride (PE-g-MA) on thermal properties, morphology, and tensile properties of low-density polyethylene (LDPE) and corn starch blends. J. Appl. Polym. Sci. **88**(13), 2904–2911 (2003). https://doi.org/10.1002/app.11965

42. Luo, F., Ning, N., Chen, L., Su, R., Cao, J., Zhang, Q., Fu, Q., Zhao, S.: Effects of compatibilizers on the mechanical properties of low density polyethylene/lignin blends. Chin. J. Polym. Sci. **27**(6), 833–842 (2009). https://doi.org/10.1142/s0256767909004552
43. Salleh, F.M., Hassan, A., Yahya, R., Lafia-Araga, R.A., Azzahari, A.D., Nazir, M.N.Z.M.: Improvement in the mechanical performance and interfacial behavior of kenaf fiber reinforced high density polyethylene composites by the addition of maleic anhydride grafted high density polyethylene. J. Polym. Res. **21**, 439 (2014). https://doi.org/10.1007/s10965-014-0439-y
44. ASTM D790-17, Standard Test Methods for Flexural Properties of Unreinforced and Reinforced Plastics and Electrical Insulating Materials, ASTM International, West Conshohocken, PA (2017)
45. ASTM E168-16, Standard Practices for General Techniques of Infrared Quantitative Analysis, ASTM International, West Conshohocken, PA (2016)
46. ASTM E1252-98(2013)e1, Standard Practice for General Techniques for Obtaining Infrared Spectra for Qualitative Analysis, ASTM International, West Conshohocken, PA (2013)
47. ASTM E2015-04(2014), Standard Guide for Preparation of Plastics and Polymeric Specimens for Microstructural Examination, ASTM International, West Conshohocken, PA (2014)
48. Rahman, M.R.: Wood Polymer Nanocomposites, 1st ed. Springer (2018). https://doi.org/10.1007/978-3-319-65735-6
49. ASTM D3418-15, Standard Test Method for Transition Temperatures and Enthalpies of Fusion and Crystallization of Polymers by Differential Scanning Calorimetry, ASTM International, West Conshohocken, PA (2015)
50. ASTM E1269-11(2018), Standard Test Method for Determining Specific Heat Capacity by Differential Scanning Calorimetry, ASTM International, West Conshohocken, PA, 2018
51. ASTM E1131-20, Standard Test Method for Compositional Analysis by Thermogravimetry, ASTM International, West Conshohocken, PA (2020)
52. Li, L., Liu, G., Zhang, C.-Y., Ou, Q.-H., Zhang, L., Zhao, X.-X.: Discrimination of bamboo using FTIR spectroscopy and statistical analysis. Guang Pu Xue Yu Guang Pu Fen Xi/Spectrosc. Spectr. Anal. **33**(12), 231–245 (2013). https://doi.org/10.3964/j.issn.1000-0593(2013)12-3221-05
53. Sharma, A.K., Chaudhary, G., Kaushal, I., Bhardwaj, U., Mishra, A.: Studies on nanocomposites of polyaniline using different substrates. Am. J. Polym. Sci. **5**(1), 1–6 (2015). https://doi.org/10.5923/s.ajps.201501.01
54. Sharma, B., Gatóo, A., Bock, M., Ramage, M.: Engineered bamboo for structural applications. Constr. Build. Mater. **81**(23), 66–73 (2015). https://doi.org/10.1016/j.conbuildmat.2015.01.077
55. Ivashchenko, O., Jurga-Stopa, J., Coy, E., Peplinska, B., Pietralik, Z., Jurga, S.: Fourier transform infrared and Raman spectroscopy studies on magnetite/Ag/antibiotic nanocomposites. Appl. Surf. Sci. **12**(1), 364–378 (2016). https://doi.org/10.1016/j.apsusc.2015.12.149
56. Islam, M.S., Kovalcik, A., Hasan, M., Thakur, V.K.: Natural fiber reinforced polymer composites. Int. J. Polym. Sci. (2015). https://doi.org/10.1155/2015/813568
57. Lu, T., Jiang, M., Jiang, Z., Hui, D., Wang, Z., Zhou, Z.: Effect of surface modification of bamboo cellulose fibers on mechanical properties of cellulose/epoxy composites. Compos. B Eng. **8**(3), 3321–3346 (2013). https://doi.org/10.1016/j.compositesb.2013.02.031
58. Hayati-Ashtiani, M.: Use of FTIR spectroscopy in the characterization of natural and treated nanostructured bentonites (montmorillonites). Part. Sci. Technol. **30**(6), 553–564 (2012). https://doi.org/10.1080/02726351.2011.615895
59. Jagtap, S.B., Mohan, M.S., Shukla, P.G.: Improved performance of microcapsules with polymer nanocomposite wall: preparation and characterization. Polym. **83**(25), 27–33 (2016). https://doi.org/10.1016/j.polymer.2015.12.011
60. Venkatesan, R., Rajeswari, N.: TiO_2 nanoparticles/poly (butylene adipate-co-terephthalate) bionanocomposite films for packaging applications. Polym. Adv. Technol. **10**(2), 40–42 (2017). https://doi.org/10.1002/pat.4042
61. Xu, Y., Guo, Z., Fang, Z., Peng, M., Shen, L.: Combination of double-modified clay and polypropylene-graft-maleic anhydride for the simultaneously improved thermal and mechanical properties of polypropylene. J. Appl. Polym. Sci. **128**(1), 283–291 (2013). https://doi.org/10.1002/app.38178

62. Tao, Y.B., You, Y., He, Y.L.: Lattice boltzmann simulation on phase change heat transfer in metal foams/paraffin composite phase change material. Appl. Therm. Eng. **93**(16), 476–485 (2016). https://doi.org/10.1016/j.applthermaleng.2015.10.016
63. Sultan, M.T., Rahman, M.R., Hamdan, S., Hossen, M.F., Mazlan, A.B.: Improved interfacial interaction between wood and styrene with the help of organically modified nanoclay. BioResources **13**(4), 8100–8112 (2018)
64. Singh, J.S.K., Ching, Y.C., Abdullah, L.C., Ching, K.Y., Razali, S., Gan, S.N.: Optimization of mechanical properties for polyoxymethylene/glass fiber/polytetra fluoroethylene composites using response surface methodology. Polym. **10**(3), 231–243 (2018). https://doi.org/10.3390/polym10030338
65. Mahato, D.N., Mathur, B.K., Bhattacherjee, S.: DSC and IR methods for determination of accessibility of cellulosic coir fibre and thermal degradation under mercerization. Indian J. Fibre Text. Res. **38**(1), 96–100 (2013)
66. Mattos, B.D., de Cademartori, P.H.G., Missio, A.L., Gatto, D.A., Magalhães, W.L.E.: Wood-polymer composites prepared by free radical in situ polymerization of methacrylate monomers into fast-growing pinewood. Wood Sci. Technol. **49**(6), 1281–1294 (2015). https://doi.org/10.1007/s00226-015-0761-5
67. Bao, S.C., Daunch, W.A., Sun, Y.H., Rinaldi, P.L., Marcinko, J.J., Phanopoulos, C.: Solid state two-dimensional NMR studies of polymeric diphenylmethane diisocyanate (PMDI) reaction in wood. For. Prod. J. **53**(6), 63–71 (2003)
68. Rahman, M.R., Islam, M.N., Huque, M.M.: Influence of fiber treatment on the mechanical and morphological properties of sawdust reinforced polypropylene composites. J. Polym. Environ. **10**(9), 23–30 (2010). https://doi.org/10.1007/s10924-010-0230-z
69. Phetkaew, W., Kyokong, B., Khongtong, S., Mekanawakul, M.: Effect of pre-treatment and heat treatment on tensile and thermal behavior of Parawood strands. Songklanakarin J. Sci. Technol. **31**(3), 323–330 (2009)
70. Liu, Z., Jiang, Z., Fei, B., Liu, X.: Thermal decomposition characteristics of Chinese fir. BioResources **8**(4), 14–24 (2013). https://doi.org/10.15376/biores.8.4.5014-5024
71. García, M., Hidalgo, J., Garmendia, I., García-Jaca, J.: Wood-plastics composites with better fire retardancy and durability performance. Compos. A Appl. Sci. Manuf. **40**(11), 1772–1776 (2009). https://doi.org/10.1016/j.compositesa.2009.08.010
72. Sánchez-Jiménez, P.E., Pérez-Maqueda, L.A., Perejón, A., Criado, J.M.: Nanoclay nucleation effect in the thermal stabilization of a polymer nanocomposite: a kinetic mechanism change. J. Phys. Chem. C **116**(21), 11797–11807 (2012). https://doi.org/10.1021/jp302466p

Acrylation and Acrylonitrile Grafting with MMT Bamboo Nanocomposite

Md Rezaur Rahman, Sinin Hamdan, and Muhammad Khusairy Bin Bakri

Abstract This chapter focuses on the impregnation modification on a grafted bamboo fiber using montmorillonite (MMT), acrylonitrile (AN) and methyl methacrylate (MMA). Two types of bamboo nanocomposites, i.e. bamboo/MMA/MMT and bamboo/AN/MMT was fabricated to enhance its mechanical, thermal and physical properties. Due to the hydrophilic nature bamboo, water in humid condition was absorbed. The present of hydroxyl group was a reason for the bamboo to have low weatherability, which prone to decomposition. Therefore, to produce bamboo nanocomposite, chemical modification is needed to create high durability fiber. As compared to the raw untreated samples, the FTIR spectrum shown less presence of cellulose, hemicellulose and pectin in the treated bamboo nanocomposites. It was observed the samples were successfully grafted with the polymer matrix containing AN/MMT or MMA/MMT. The EDS analysis showed that both untreated and treated samples were having significant amount of carbon, oxygen, followed by nitrogen, chlorine, potassium and aluminum. While the SEM shows that treated bamboo nanocomposites had better dispersion, good interfacial relation, smooth surface, less void and agglomeration. As compared to the untreated bamboo nanocomposites, it was also justified that the treated bamboo nanocomposites were having better MOE and MOR.

Keywords Bamboo · Acrylonitrile · Montmorillonite · Nanocomposites · Optimization

1 Introduction

Fiber can be extracted from plants, mineral or animals. Due to the vast development and higher awareness among researchers and industries, most of them are taking big initiatives to develop more technologies, which can sustain towards degradability, eco-friendly material properties. Furthermore, most of the synthetic fiber was very

M. R. Rahman (✉) · S. Hamdan · M. K. B. Bakri
Faculty of Engineering, Universiti Malaysia Sarawak, Jalan Datuk Mohammad Musa, 94300, Kota Samarahan, Sarawak, Malaysia
e-mail: rmrezaur@unimas.my

M. R. Rahman (ed.), *Bamboo Polymer Nanocomposites*, Engineering Materials,
https://doi.org/10.1007/978-3-030-68090-9_3

expensive and non-degradable, which made natural fiber as a better choice, even though it is not durable enough for some high-end engineering applications. In addition, the natural fiber drawbacks could be decreased by reinforcing the nano clay into the fiber, which produce its nanocomposite and create enhance physical and mechanical properties. One of the good examples is as a source for the natural fiber.

Bamboo is categorized under the group of grass family, which is also the fastest growing woody plants in the world and reportedly grow up to almost hundred inches per day [1]. The high bamboo growth rate makes it versatile, sustainable and environmentally friendly as natural resources for the future research. Although it is less durable as compared to the modified bamboo and conventional materials, i.e. wood and steel, the fresh bamboo readily had good mechanical properties, cheap and safe [2]. Moreover, as compared to steel and low modulus of elasticity, the untreated bamboo is susceptible to environmental degradation due to low weatherability, biological attack, shorter life-year span [3]. Therefore, to enhance the bamboo physical and mechanical properties, several modification methods are introduced by researchers all over the world due to this issue [4].

Grafted hydrophobic matrix such as acrylonitrile (AN) and methyl methacrylate (MMA) is aimed to form branched copolymer. As one side chain of the graft is covalently bonded to the main chain of a polymer backbone, which altering the physical and chemical properties of targeted polymer without affecting its original properties [5, 6]. Moreover, depending on the type of the matrix used, the natural polymer modification acquired novel properties, such as hydrophobic nature, thermal resistivity and biological resistivity [7]. To focus on the bamboo enhancement, the modification is expected by means of impregnation technique and the optimum fabrication conditions was determined by using respond surface methodology. The treated bamboo nanocomposite was characterized mechanically and physically and compared to the conventional materials to analyze the differences in their properties. Based on Zaki and Abdullah [8], the morphology studies have proved that the diameter and the surface roughness of grafted polymer are increased with better physical and mechanical properties. Therefore, this chapter discussed the results of the bamboo nanocomposites through characterization, i.e. physico-chemical morphological analysis, and mechanical properties. The degree of changes based on the result are shown in the modification process, which were optimized based on AN, MMT and MMA.

2 Methodology

2.1 Materials

The bamboo (*Gigantochloa scortechenii*) was obtained from a forest in Kota Samarahan, Sarawak, Malaysia. Analytical grade chemicals, including methyl methacrylate (MMA), acrylonitrile (AN) and montmorillonite (MMT) (99% purity, Sigma-Aldrich, St. Louis, MO, USA), ethanol and benzoyl peroxide (95% purity, Braun HmbG, Kronberg, Germany) were also used.

2.2 Bamboo Sample Preparation

Raw bamboos were obtained from a forest in Kota Samarahan, Sarawak, Malaysia. Figure 1 shown the bamboos that were cut into few strips with dimension of approximately 0.5 × 2.0 × 35 cm (thickness, width, and length). To produce treated bamboo nanocomposite, the bamboo strips were dried for 5 days in an oven at 70°C, which helps remove the moisture before proceeding with impregnation of the bamboo strips. Analytical grade chemicals were used includes acrylonitrile (AN), methyl methacrylate (MMA) and montmorillonite nano clay (MMT). As catalyst, to speed up the

Fig. 1 Bamboo strips

reaction mixture, benzoyl peroxide was used in polymer matrix. While to dissolve the radicals, ethanol was used during polymerization of the samples.

2.3 Preparation of Bamboo/MMA/MMT Nanocomposite and Bamboo/AN/MMT Nanocomposite

For bamboo/MMA/MMT nanocomposites, the MMT and MMA were dissolved in 500 mL ethanol, while for the bamboo/AN/MMT nanocomposites, the MMT, AN and MMA were dissolved in 500 mL ethanol. To act as catalyst, 5 g of benzoyl peroxide was added into the matrix solution. The matrix solution was shown in Fig. 2. A dried bamboo strip and the matrix solution were placed in the vacuum impregnator that was connected to an air compressor. The bamboo strip must drown in the matrix solution. Hence, heavy load was placed on top of the bamboo strip to dip into the solution. Figure 3 shows the lid of the vacuum impregnator was closed and locked tightly, and a slightly above atmospheric pressure was exerted into the chamber by an air compressor to impregnate the sample for about 30 min as shown in Fig. 4. The impregnation was done in a fume chamber to release the toxic vapors. After 30 min, the sample was taking out from the vacuum impregnator. The sample was rinse with distilled water and wiped with dry cloth to remove excess moisture. Finally, the sample was wrapped up with aluminum foil and was dried for 24 h at 80°C in an oven. The steps were repeated for other samples.

Fig. 2 Matrix solution

Fig. 3 Tightly closed vacuum impregnator

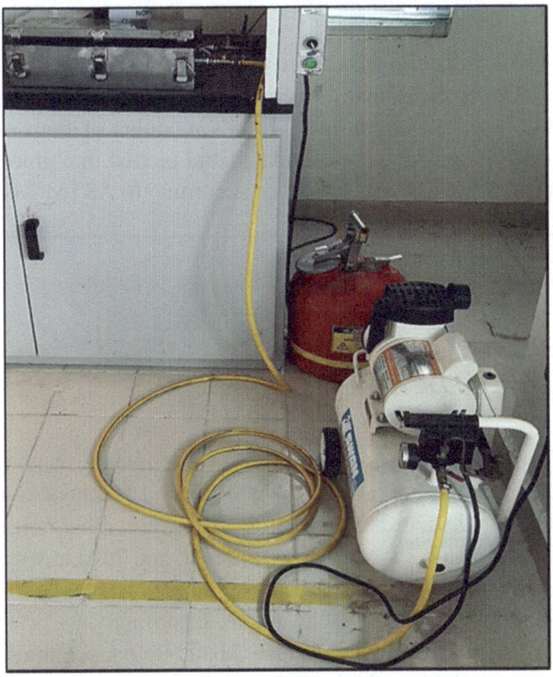

Fig. 4 Vacuum impregnator connected to an air compressor

2.4 Fourier Transform Infrared Spectroscopy (FTIR)

The FTIR spectrum of the untreated and treated bamboo nanocomposites consist of raw bamboo, bamboo/MMA/MMT and bamboo/AN/MMT nanocomposites were measured, recorded and plotted in the range of 4000 to 400 cm^{-1} wavenumber absorption bands in spectrophotometer. The test was done according to ASTM E168-16 [9], and ASTM E1252-98 [10] standards.

2.5 Scanning Electron Microscopy (SEM) and Energy Dispersive X-Ray Spectroscopy (EDS)

The magnified SEM images of untreated and treated bamboo nanocomposites (Bamboo/MMA/MMT and Bamboo/AN/MMT) were obtained the surface morphological of both untreated and treated bamboo nanocomposites. Both tests were conducted according to ASTM E2015-04 [11] and ASTM E1508-12 [12] standards. The EDS was repeated multiple times to get the most repeated results.

2.6 Tensile Test

The tensile tests were carried out using Universal Testing Machine Shimadzu MSC-5/500 (Kyoto, Japan), operated at a speed of 5 mm/min. The generated results for the tensile strength and Young's modulus of both untreated and treated bamboo nanocomposite were calculated and plotted into graphs to analyses the mechanical properties. The test was done according to ASTM D638-14 [13].

3 Results and Discussions

3.1 FTIR Analysis

The FTIR spectrum of untreated and treated bamboo nanocomposites consist of raw bamboo, bamboo/MMA/MMT and bamboo/AN/MMT nanocomposites were measured, recorded and plotted in a wavenumber range of 4000 to 400 cm^{-1} absorption bands as shown in Figs. 5, 6a–d, and 7a–d, respectively in order to identify the chemical bonds presence in the samples. Tables 1, 2 and 3 tabulated the respective functional group of each sample, accordingly.

Figure 5 show the untreated bamboo nanocomposite FTIR analysis. According to Li et al. [14], the bamboo fibers main constituents was cellulose, hemicellulose and lignin. Based on Rana et al. [15], the bamboo cellulose contained hydroxyl group, which was responsible for the bamboo fibers inherent hydrophilic nature. This indicated that the O-H stretching happen at the wavenumber of 3336.03 cm^{-1}, and 2333.97 cm^{-1}, and C-OH stretch at 1031.96 cm^{-1}. Figure 5 shows that the cellulose was designated by the O-H stretching at wavenumber between 3257.77 to 2441.88 cm^{-1}, and C-OH stretching at 1028 cm^{-1}. According to Rahman et al.

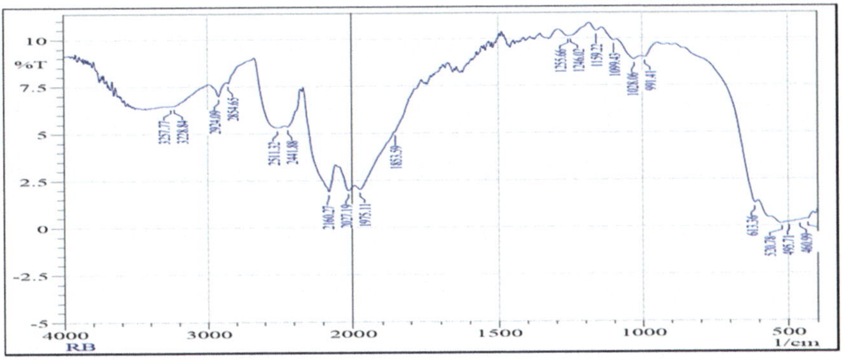

Fig. 5 Untreated bamboo

[16], in bamboo fibers, the C-H stretching ranges from 3000 to 2850 cm^{-1} was indicated for the presence of methyl and methylene groups. Figure 5 shows that the C-H stretching in methyl and methylene groups are ranging from 2924.09 to 2854.65 cm^{-1}. Furthermore, the 1255.66 cm^{-1} peak intensities wavenumber was assigned to the C-O stretching vibration of acetyl group, in hemicelluloses and lignin. This was validated by a research conducted by Sinha and Rout [17], whereas the C-O stretching of the acetyl group is around 1259 cm^{-1}.

The FTIR analysis of bamboo/MMA/MMT nanocomposites are shown in Fig. 6a–d, respectively. It shows that the bamboo nanocomposite was treated with MMA and MMT to modify the physical and chemical properties of the bamboo. Figure 6a shows the FTIR analysis of treated sample grafted with 10wt% MMA and 5% MMT. It was observed that the significant O-H stretching of the MMA/MMT treated bamboo nanocomposite is ranging from 3747.69 to 2441.88 cm^{-1}. Between 4000 and 2500 cm^{-1}, the O-H stretching was the same trend for the rest of the MMA/MMT treated samples, as shown in Fig. 6b–d. A much higher intensity peak from the range of 1750 to 1735 cm^{-1} was associated with $C = O$ stretching of esters group, which prove that the MMA was successfully grafted into bamboo nanocomposite. Thakur et al. [18] stated that the new peak occurrence at that band was associated with the presence of the grafted MMA into fibers.

Figure 7a–d respectively shows the bamboo/AN/MMT nanocomposites FTIR analysis. Modified and treated bamboo nanocomposite was produced by introduced ANs and MMTs into the fibers of bamboo. From all the AN/MMT treated samples FTIR analysis in Fig. 7a–d, the 3500 to 2400 cm^{-1} bands was corresponded to the vibration of hydrogen bond and O-H groups. While the 2930 to 2800 cm^{-1} bands was responsible for the C-H stretching. Based on Dahou et al. [5], the AN presence in a sample is observed at 2260 to 2222 cm^{-1} band was corresponded to the CN vibration of nitrile group. This means that the AN grafted into the samples are successful, if bands are within those ranges. Furthermore, the MMT impregnated into bamboo nanocomposite for both acrylation and acrylonitrile grafting was seen on the peak intensities at approximately 3500 to 3000 cm^{-1}, 1050 to 1020 cm^{-1} and

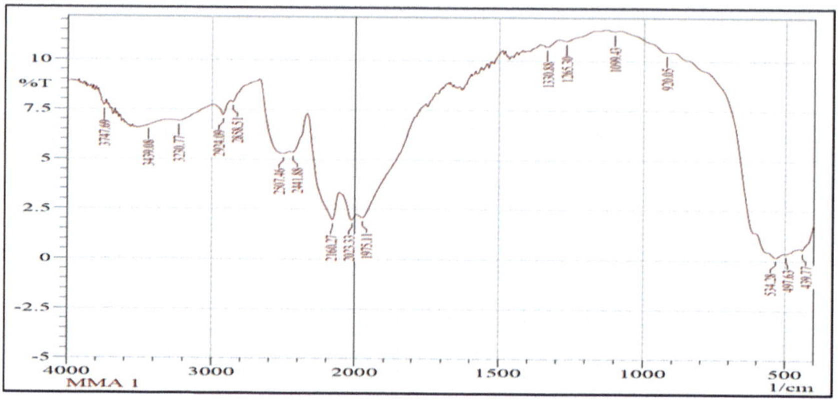

a

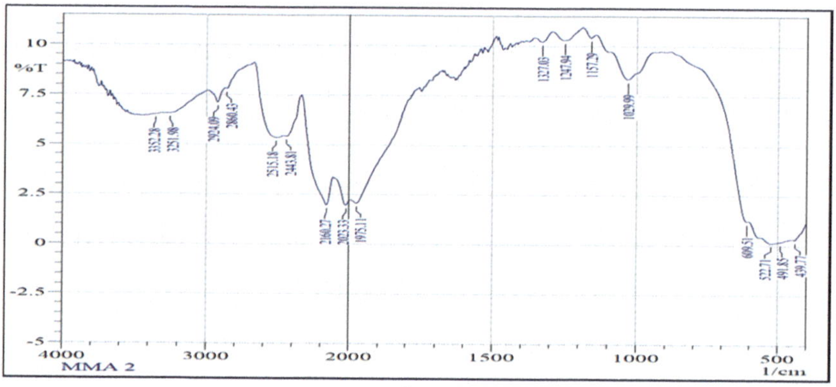

b

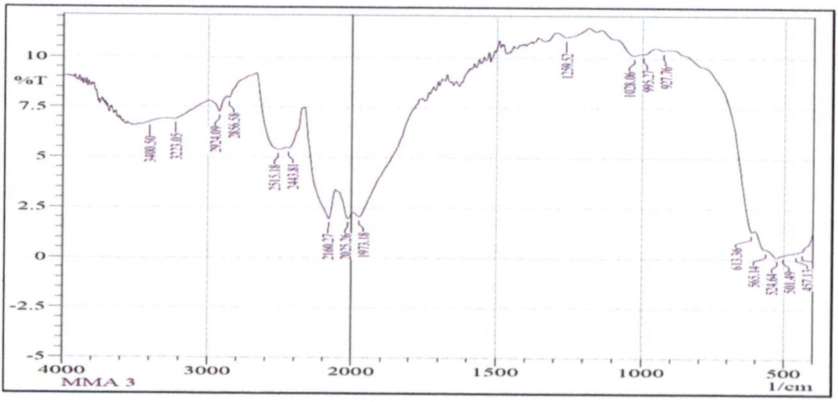

c

Fig. 6 **a** Treated bamboo with 10 wt% MMA and 5 wt% MMT. **b** Treated bamboo with 1 wt% MMA and 25 wt% MMT. **c** Treated bamboo with 10 wt% MMA and 25 wt% MMT. **d** Treated bamboo with 1 wt% MMA and 5 wt% MMT

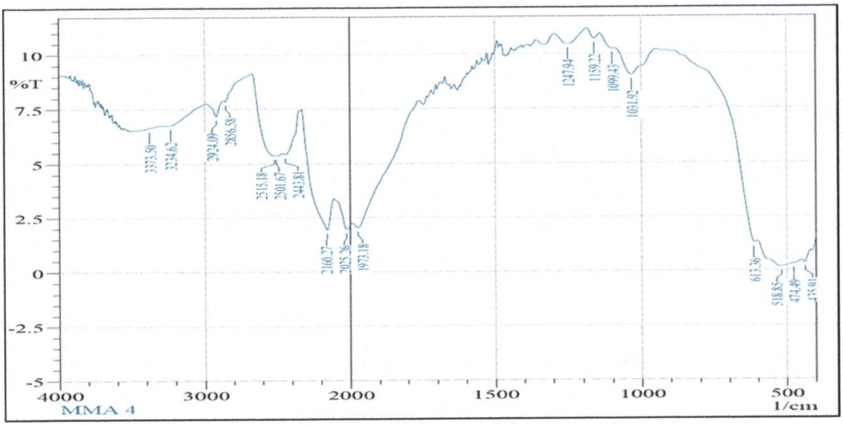

d

Fig. 6 (continued)

670 to 500 cm^{-1} in all MMA/MMT and AN/MMT treated bamboo samples. This was agreed by Jagtap et al. [19], whereas the peak intensities were ranges similarly with the grafted MMT.

3.2 *Energy Dispersive X-Ray Spectroscopy (EDS) Analysis*

EDS analysis was used to analyses the elemental composition presence in both untreated and treated bamboo nanocomposites. The results obtained are analyzed, tabulated, and discussed accordingly.

Figure 8a–c shows the EDS analysis of the raw untreated bamboo nanocomposite, AN/MMT treated bamboo nanocomposite, and MMA/MMT treated bamboo nanocomposite. Table 4 was the tabulated based on the elemental composition presence in the samples. Both Fig. 8a–c and Table 4 (a) to (c) was deduced from both untreated and treated bamboo nanocomposites spectrum. The presence of the major elements such as carbon (C) and oxygen (O) was possibly originated from the bamboo cellulose, whereas 74% of bamboo's constituents are made of cellulose as stated by Li et al. [14]. Moreover, C and O was originated from $C=O$ stretching in the chemical treatment of MMA/MMT, which presented in the FTIR analysis. This indicated the presence of ester group in MMA/MMT treated bamboo nanocomposite. Furthermore, the nitrogen (N) in AN/MMT treated bamboo nanocomposite was believed coming from the nitrile group of AN. Potassium (K), aluminum (Al) and chlorine (Cl) traces was shown in the EDS analysis due to the materials reaction or impurities traces in bamboo.

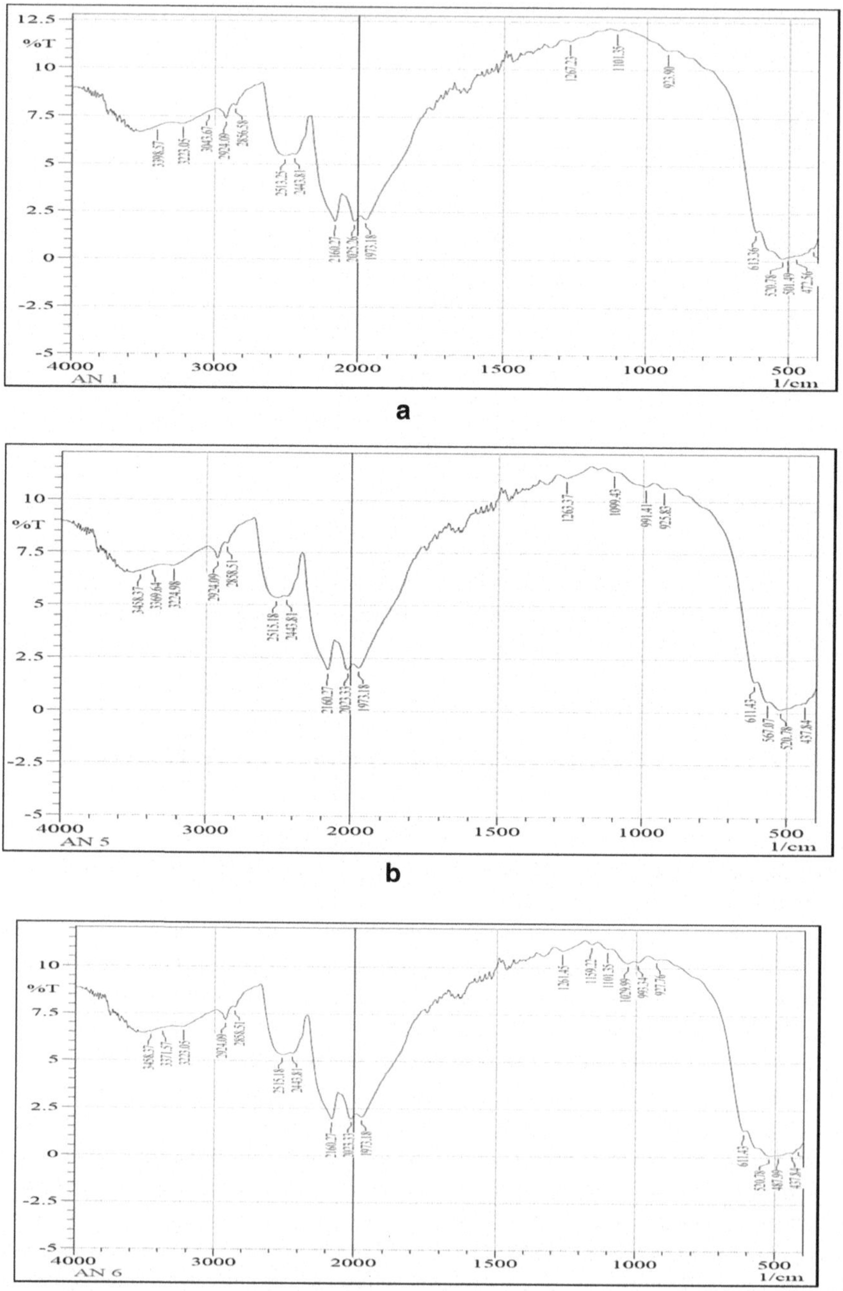

Fig. 7 **a** Treated bamboo with 1 wt% AN and 5 wt% MMT. **b** Treated bamboo with 10 wt% AN and 5 wt% MMT. **c** Treated bamboo with 1 wt% AN and 25 wt% MMT. **d** Treated bamboo with 10 wt% AN and 25 wt% MMT

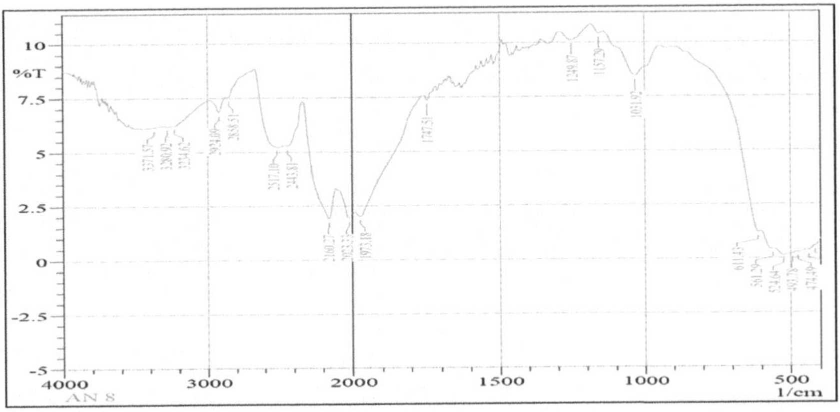

d

Fig. 7 (continued)

Table 1 Functional groups present in untreated bamboo

Peak	Group frequency (cm^{-1})	Functional group
1	3257.77	O-H stretch
2	3228.84	O-H stretch
3	2924.09	O-H stretch, C-H stretch
4	2854.65	O-H stretch, C-H stretch
5	2511.32	O-H stretch
6	2441.88	O-H stretch
7	2160.27	S-C ≡ N stretch
8	2027.19	N = C=S stretch
9	1975.11	C-H bend
10	1853.59	C-H bend
11	1255.66	C-O Stretch
12	1246.02	C-O-C stretch
13	1159.22	C-OH stretch
14	1099.43	C-OH stretch
15	1028.06	C-OH stretch
16	991.41	C = C bend

3.3 Scanning Electron Microscopy (SEM) Analysis

The SEM images of untreated and treated bamboo nanocomposites (Bamboo/MMA/MMT and Bamboo/AN/MMT) were obtained to analyze the surface morphological of both untreated and treated bamboo nanocomposites, as shown in Figs. 9, 10a–d, and 11a–d.

Table 2 Functional group present in bamboo/MMA/MMT nanocomposite

Sample	Peak	Group frequency (cm^{-1})	Functional group
6(a)	1	3747.69	O-H stretch
	2	3439.08	O-H stretch
	3	3230.77	O-H stretch
	4	2924.09	O-H stretch, C-H stretch
	5	2858.51	O-H stretch, C-H stretch
	6	2507.46	O-H stretch
	7	2441.88	O-H stretch
	8	2160.27	S-C ≡ N stretch
	9	2023.33	N = C=S stretch
	10	1975.11	C-H bend
	11	1330.88	O-H bend
	12	1265.30	C-O stretch
	13	1099.43	C-OH stretch
6(b)	1	3352.28	O-H stretch
	2	3251.98	O-H stretch
	3	2924.09	O-H stretch, C-H stretch
	4	2860.43	O-H stretch, C-H stretch
	5	2515.18	O-H stretch
	6	2443.81	O-H stretch
	7	2160.27	S-C ≡ N stretch
	8	2023.33	N = C=S stretch
	9	1975.11	C-H bend
	10	1327.03	O-H bend
	11	1247.94	C-O-C stretch
	12	1157.29	C-OH stretch
	13	1029.99	C-OH stretch
6(c)	1	3400.50	O-H stretch
	2	3223.05	O-H stretch
	3	2924.09	O-H stretch, C-H stretch
	4	2856.58	O-H stretch, C-H stretch
	5	2515.18	O-H stretch
	6	2443.81	O-H stretch

(continued)

Table 2 (continued)

Sample	Peak	Group frequency (cm^{-1})	Functional group
	7	2160.27	S-C ≡ N stretch
	8	2025.26	N = C=S stretch
	9	1973.18	C-H bend
	10	1259.52	C-O stretch
	11	1028.06	C-OH stretch
	12	995.27	C = C bend
6(d)	1	3373.50	O-H stretch
	2	3234.62	O-H stretch
	3	2924.09	O-H stretch, C-H stretch
	4	2856.58	O-H stretch, C-H stretch
	5	2515.18	O-H stretch
	6	2501.67	O-H stretch
	7	2443.81	O-H stretch
	8	2160.27	S-C ≡ N stretch
	9	2025.26	N = C=S stretch
	10	1973.18	C-H bend
	11	1247.94	C-O-C stretch
	12	1159.22	C-OH stretch
	13	1099.43	C-OH stretch
	14	1031.92	C-OH stretch

Figure 9 shows the SEM image of untreated bamboo. The untreated bamboo nanocomposite shows traced of withdrawal of fibers, jaggy surfaces, agglomeration and cavities. This situation was due to the weak interfacial bonding and poor fiber distribution in the matrix, as reported by Islam et al. [20]. It was observed that debonding between polymer and fiber are also observed for the untreated bamboo nanocomposite. There are numerous voids were observed on the fracture surface, as no matrix stick to the untreated bamboo. According to Bonnia et al. [21], the lack interaction at the surface and the effect of agglomerations was due to debonding, which taking place in the fibers. It was also observed that the untreated bamboo nanocomposite having impurities on its surface. These impurities caused weak interfacial bonding between polymer and matrix. According to Azwa and Yousif [22], more spaces allowed the moisture to occupy the weak interfacial bonding.

Figure 10a–d shows the SEM images of treated bamboo nanocomposites with MMA and MMT at different weight percentage. It was observed almost no difference in interfacial interaction between fiber and polymer matrix in the nanocomposite

Table 3 Functional group present in bamboo/AN/MMT nanocomposite

Sample	Peak	Group frequency (cm^{-1})	Functional group
7(a)	1	3398.57	O-H stretch
	2	3223.05	O-H stretch
	3	3043.67	O-H stretch, C-H stretch
	4	2924.09	O-H stretch, C-H stretch
	5	2856.58	O-H stretch, C-H stretch
	6	2513.25	O-H stretch
	7	2443.81	O-H stretch
	8	2160.27	S-C ≡ N stretch
	9	2025.26	N = C=S stretch
	10	1973.18	C-H bend
	11	1267.23	C-O stretch
	12	1101.35	C-OH stretch
7(b)	1	3458.37	O-H stretch
	2	3369.64	O-H stretch
	3	3224.98	O-H stretch
	4	2924.09	O-H stretch, C-H stretch
	5	2858.51	O-H stretch, C-H stretch
	6	2515.18	O-H stretch
	7	2443.81	O-H stretch
	8	2160.27	S-C ≡ N stretch
	9	2023.33	N = C=S stretch
	10	1973.18	C-H bend
	11	1263.37	C-O stretch
	12	1099.43	C-OH stretch
	13	991.41	C = C bend
7(c)	1	3458.37	O-H stretch
	2	3371.57	O-H stretch
	3	3223.05	O-H stretch
	4	2924.09	O-H stretch, C-H stretch
	5	2858.51	O-H stretch, C-H stretch
	6	2515.18	O-H stretch

(continued)

Table 3 (continued)

Sample	Peak	Group frequency (cm^{-1})	Functional group
	7	2443.81	O-H stretch
	8	2160.27	S-C ≡ N stretch
	9	2023.33	N = C=S stretch
	10	1973.11	C-H bend
	11	1261.45	C-O stretch
	12	1159.22	C-OH stretch
	13	1101.35	C-OH stretch
	14	1029.99	C-OH stretch
	15	993.34	C = C bend
7(d)	1	3371.57	O-H stretch
	2	3280.92	O-H stretch, C-H stretch
	3	3234.62	O-H stretch
	4	2924.09	O-H stretch, C-H stretch
	5	2858.51	O-H stretch, C-H stretch
	6	2517.10	O-H stretch
	7	2443.81	O-H stretch
	8	2160.27	S-C ≡ N stretch
	9	2023.33	N = C=S stretch
	10	1973.18	C-H bend
	11	1747.51	C-H bend, C = O stretch
	12	1249.87	C-O-C stretch
	13	1157.29	C-OH stretch
	14	1031.92	C-OH stretch

systems. From Fig. 10a, proper attachment between fiber and polymer matrix indicated interfacial bonding between them are strong. It observed that there was better dispersion, good interfacial bonding, less cavity and agglomeration displayed in the SEM image of 10 wt% MMA and 5wt% MMT. The increase in interfacial adhesion between fiber and polymer matrix created less voids at the end of fiber pull-out [23]. Figure 10b–d shows that there was lot of impurities presence in the surface of the samples, which might due to the failure in alkaline treatment as reported by Lin et al. [24]. Moreover, SEM images of Fig. 10b–d shown less fiber pull-out. According to Ray et al. [25], less pull-out of fibers from the fracture point is the result of good interfacial adhesion between fiber and polymer matrix during fiber treatment.

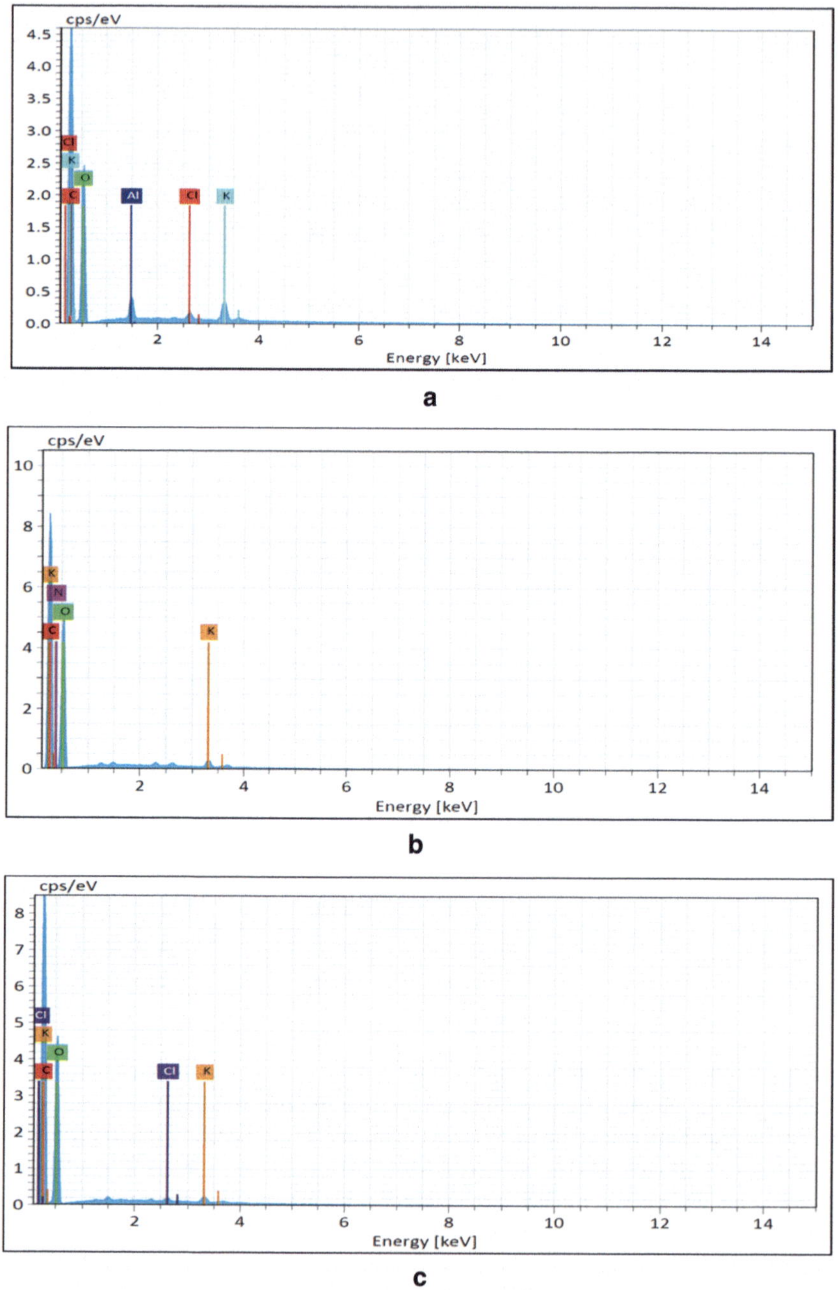

Fig. 8 **a** EDS analysis of raw untreated bamboo nanocomposite. **b** EDS analysis of AN/MMT treated bamboo nanocomposite. **c** EDS analysis of MMA/MMT treated bamboo nanocomposite

Table 4 Elemental composition of untreated and treated samples

Samples	Elements	Atom (%)
4.4(a)	C	61.11
	O	37.60
	K	0.68
	Al	0.47
	Cl	0.13
4.4(b)	C	56.97
	O	41.02
	N	1.71
	K	0.30
4.4(c)	C	61.20
	O	38.50
	K	0.21
	Cl	0.08

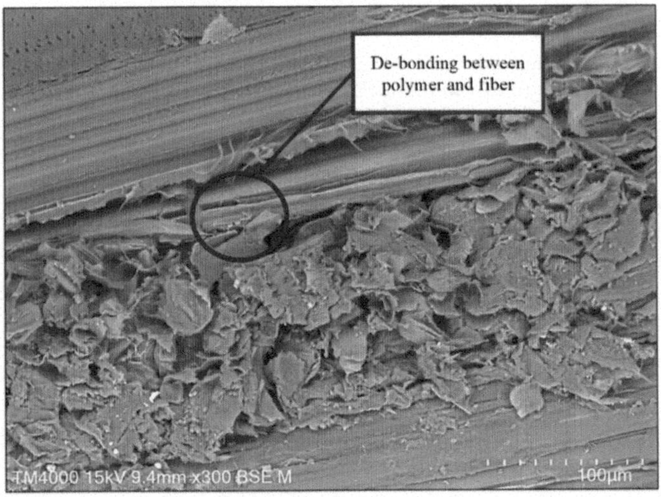

Fig. 9 SEM image of untreated bamboo

Figure 11a–d shows the SEM images of treated bamboo nanocomposites with AN and MMT with different weight percentage. It was observed there was lot of impurities presence in the surface of the treated samples. This was also due to the failure in alkaline treatment. Figure 11a, it was deduced that the fiber-pull out is lower than that of in Fig. 11b–d. This indicated the interfacial bonding between the fibers and polymer are better due to better dispersion, less void and agglomeration.

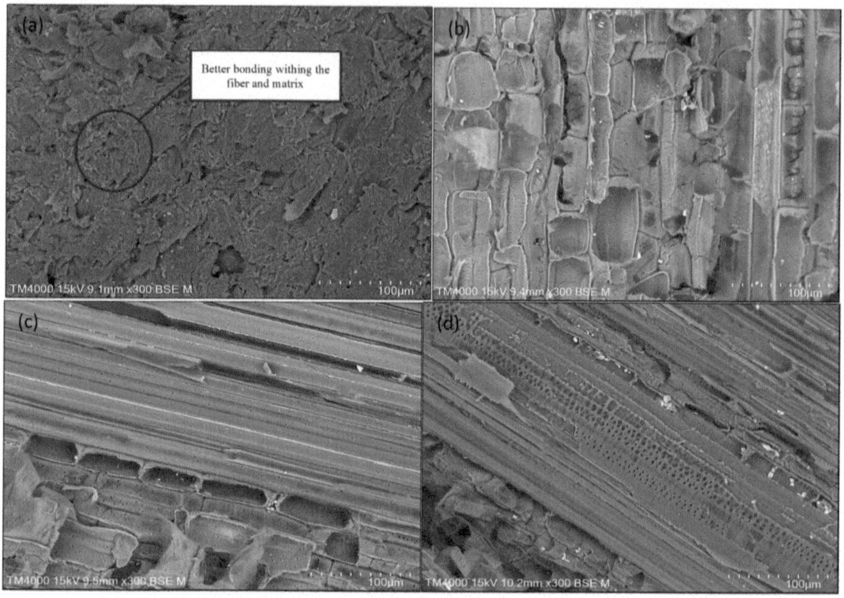

Fig. 10 SEM image of treated bamboo with, **a** 10 wt% MMA and 5 wt% MMT, **b** 1 wt% MMA and 25 wt% MMT, **c** with 10 wt% MMA and 25 wt% MMT, and **d** 1 wt% MMA and 5 wt% MMT

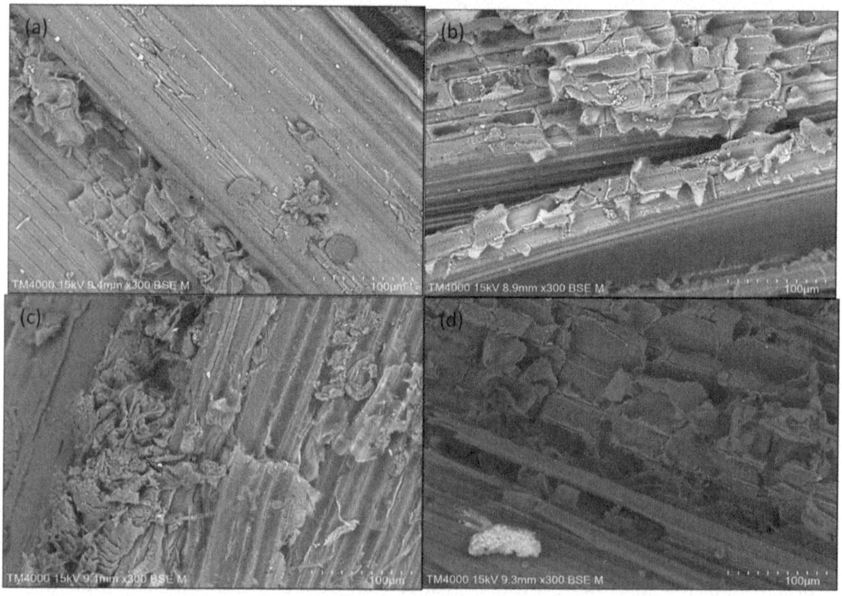

Fig. 11 SEM image of treated bamboo with **a** 1 wt% AN and 5 wt% MMT, **b** 10 wt% AN and 5 wt% MMT, **c** 1 wt% AN and 25 wt% MMT, and **d** 10 wt% AN and 25 wt% MMT

Table 5 Tensile strength and tensile modulus of the samples

Samples	Tensile strength (MPa)	Tensile modulus (GPa)
RB	34.563	1.436
MMA1	134.141	1.371
MMA2	103.662	1.247
MMA3	120.167	1.336
MMA4	119.726	1.345
AN1	149.500	1.640
AN2	136.500	1.587
AN3	118.750	1.402
AN4	155.313	1.580

3.4 *Tensile Properties*

Both untreated and treated nanocomposites consist of raw bamboo, bamboo/MMA/MMT and bamboo/AN/MMT were tested for its tensile strength and tensile modulus. The data obtained were calculated and the graphs of tensile strength and tensile modulus were plotted and discussed.

The tensile strength and tensile modulus of the untreated and treated bamboo nanocomposites are tabulated in Table 5. All samples data obtained are plotted in two in Figs. 12 and 13, for both tensile strength and Young's modulus of the samples, respectively. From Fig. 12, the treated bamboo nanocomposites tensile strength was

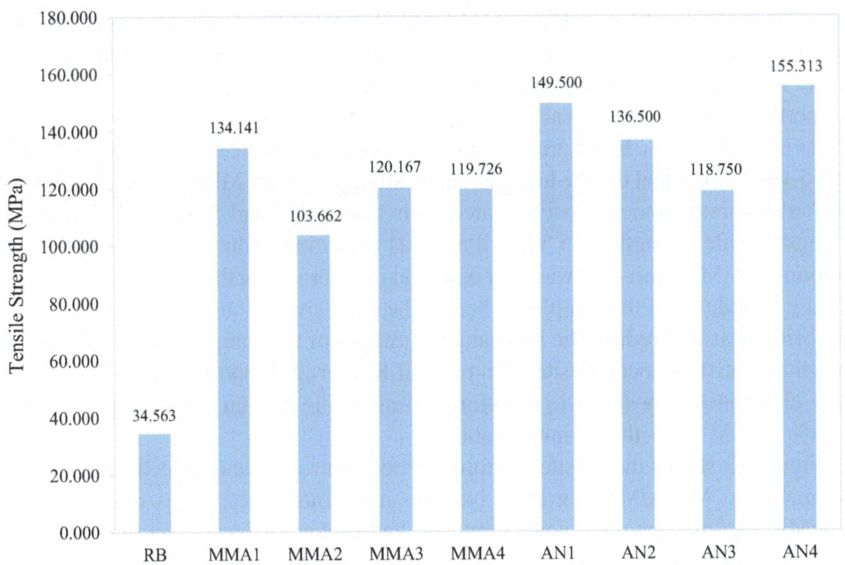

Fig. 12 Tensile strength of the samples

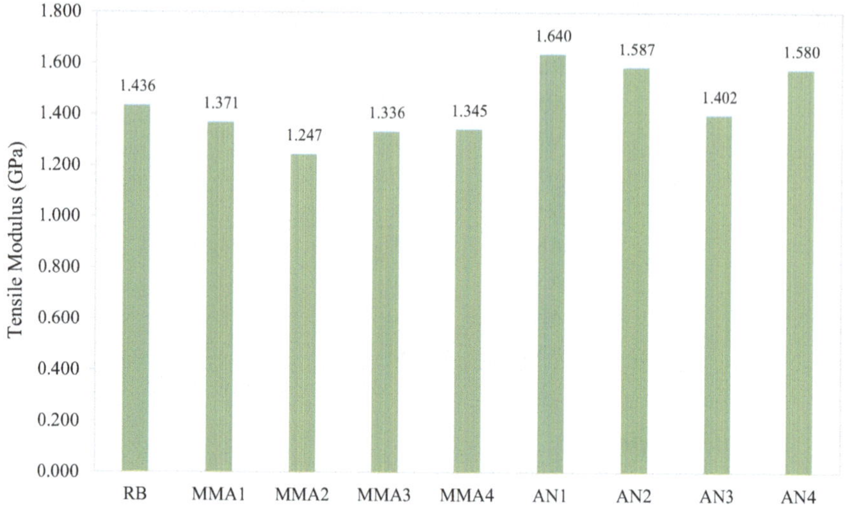

Fig. 13 Tensile modulus of the samples

observed to be higher than the untreated bamboo nanocomposite. For both acrylation and acrylonitrile, the treated samples tensile strength was greater than 100 MPa. Based on Hossen et al. [26], the treated samples mechanical properties increase due to the reduction of hydroxyl groups in the fibers, as it reacts with the chemical agents, causing stronger interfacial bonding between the fibers and polymer matrix. It was also observed that higher tensile strength of AN/MMT grafted bamboo nanocomposite was observed compared to MMA/MMT grafted bamboo nanocomposite. To enhance the mechanical properties of the samples, it concluded that AN/MMT was better than MMA/MMT as polymer matrix. Bonnia et al. [21] mentioned the tensile properties improvement was achieved as composites have good interaction and better stress transfer. MMA/MMT grated bamboo nanocomposite treated with 1 wt% MMA and 5 wt% MMT had the the highest tensile strength (149.5 MPa). While AN/MMT grafted bamboo nanocomposite treated with 1 wt% AN and 5 wt% MMT had the highest tensile strength (155.313 MPa). This condition deduced that the optimum amount of MMT nano-clay was very essential in determining the tensile strength and Young's modulus of the samples, whereas the excessive amount of MMT increased the brittleness and reduce the mechanical strength of the samples. In preparing the polymer matrix nanocomposites, Kumar and Kumar [27] reported that 5 wt% to 25 wt% clay nanocomposites was used for grafting, while Adamu et al. [28] proved that 5 wt% of MMT was the optimum value.

Figure 13 shows the tensile modulus of both untreated and treated samples. As compared to MMA/MMT grafted bamboo nanocomposite, it was observed that AN/MMT grafted bamboo nanocomposite had higher stiffness. The highest value belongs to the treated 10 wt% AN and 5 wt% MMT bamboo nanocomposites and the lowest value goes to the treated 1 wt% MMA and 25 wt% MMT bamboo nanocomposites. It also observed that the raw bamboo Young's modulus was high as compared to the treated bamboo nanocomposite. MMA/MMT grafted deduced that the presence of MMT, which was not necessarily increase the mechanical strength of the bamboo nanocomposite. However, it induced even more brittleness in the bamboo nanocomposite, due to over amount of nano-clay is used. It also causes the bamboo nanocomposite to be less elastic and more brittle as reported by Adamu et al. [28]. Therefore, it is concluded that from these tensile properties, the grafted AN/MMT grafting was better than grafted MMA/MMT due to better capability to be used as reinforce for the bamboo nanocomposite, that enhanced both tensile strength and tensile modulus of the samples.

4 Conclusions

Throughout this chapter, the FTIR proved the presence of cellulose, hemicellulose and pectin, which was clearly significant in the untreated bamboo nanocomposite. It was also observed that treated MMA/MMT and AN/MMT bamboo nanocomposite have less significant amount of bamboo's constituents due to removal lignin, pectin and wax. The presence of the grafted MMA, AN and MMT proved by the FTIR spectrum at specific wavenumbers. Besides, in the bamboo nanocomposite, the EDS proved that the presence of carbon and oxygen are associated with the presence of cellulose. The SEM on the other hand demonstrated that treated bamboo nanocomposites had better dispersion, good interfacial relation, smooth surface, less void and agglomeration as compared with the raw untreated bamboo nanocomposite. It also observed that there were still lot of impurities on the samples, which was due to no treatment was done on the untreated bamboo. Moreover, it was deduced that the tensile properties of treated bamboo nanocomposites were superior than untreated bamboo nanocomposite. Nevertheless, excessive amount of MMTs cause some of the treated bamboo nanocomposite to be brittle and having lower elasticity, as compared to raw untreated bamboo nanocomposite and some samples with optimum amount of 5 wt% MMT. From the research, in term of mechanical strength, it was concluded that the grafted AN was better than grafted MMA because of the better capability to reinforce the bamboo nanocomposite showed by both tensile strength and Young's modulus.

Acknowledgements The authors would like to acknowledge Universiti Malaysia Sarawak (UNIMAS) for the support.

References

1. Yeasmin, L., Ali, M.N., Gantait, S., Chakraborty, S.: The bamboo which categorizes under the group of grass family is a fastest growing woody plants in the world, reportedly grow up to almost hundred inches per day. Biotech **5**(1), 1–11 (2015). https://doi.org/10.1007/s13205-014-0201-5
2. Chaowana, P.: Bamboo: an alternative raw material for wood and wood-based composites. J. Mater. Sci. Res. **2**(2), 90–102 (2013). https://doi.org/10.5539/jmsr.v2n2p90
3. Mohammed, L., Ansari, M.N.M., Pua, G., Jawaid, M., Islam, M.S.: A review on natural fiber reinforced polymer composite and its applications. Int. J. Polym. Sci. **1**, 1–15 (2015). https://doi.org/10.1155/2015/24
4. Zhang, X., Wang, F., Keer, L.M.: Influence of surface modification on the microstructure and thermo-mechanical properties of bamboo fibers. Materials **8**(10), 6597–6608 (2015). https://doi.org/10.3390/ma8105327
5. Dahou, W., Ghemati, D., Oudia, A., Aliouche, D.: Preparation and biological characterization of cellulose graft copolymers. Biochem. Eng. J. **48**(2), 187–194 (2010). https://doi.org/10.1016/j.bej.2009.10.006
6. Kumar, V., Naithani, S., Pandey, D.: Optimization of reaction conditions for grafting of α-cellulose isolated from Lantana camara with acrylamide. Carbohyd. Polym. **86**(2), 760–768 (2011). https://doi.org/10.1016/j.carbpol.2011.05.019
7. Vieira, M.G.A., Silva, M.A., Santos, L.O., Beppu, M.M.: Natural-based plasticizers and biopolymer films: a review. Eur. Polymer J. **47**(3), 254–263 (2011). https://doi.org/10.1016/j.eurpolymj.2010.12.011
8. Zaki, A.F., Abdullah, I.: Graft copolymerization of acrylonitrile onto torch ginger cellulose. Sains Malaysiana **44**(6), 853–859 (2015). https://doi.org/10.17576/jsm-2015-4406-11
9. ASTM E168-16: Standard Practices for General Techniques of Infrared Quantitative Analysis. ASTM International, West Conshohocken, PA (2016). https://doi.org/10.1007/10.1520/E0168-16
10. ASTM E1252-98: Standard Practice for General Techniques for Obtaining Infrared Spectra for Qualitative Analysis. ASTM International, West Conshohocken, PA (2013). https://doi.org/10.1520/E1252-98R13E01
11. ASTM E2015-04: Standard Guide for Preparation of Plastics and Polymeric Specimens for Microstructural Examination. ASTM International, West Conshohocken, PA (2014). https://doi.org/10.1520/E2015-04R14
12. ASTM E1508-12: Standard Guide for Quantitative Analysis by Energy-Dispersive Spectroscopy. ASTM International, West Conshohocken, PA (2019). https://doi.org/10.1520/E1508-12AR19
13. ASTM D638-14: Standard Test Method for Tensile Properties of Plastics. ASTM International, West Conshohocken, PA (2014). https://doi.org/10.1520/D0638-14
14. Li, L.J., Wang, Y.P., Wang, G., Cheng, H.T., Han, X.J.: Evaluation of properties of natural bamboo fiber for application in summer textiles. J. Fiber Bioeng. Inform. **3**(2), 94–99 (2010). https://doi.org/10.3993/jfbi09201006
15. Rana, A.K., Mandal, A., Mitra, B.C., Jacobson, R., Rowell, R., Banerjee, A.N.: Short jute fibre reinforced polypropylene composites: effect of compatibilizer. J. Appl. Polym. Sci. **63**(2), 329–338 (1998). https://doi.org/10.1002/(SICI)1097-4628(19980711)69:2%3c329:AID-APP14%3e3.0.CO;2-R
16. Rahman, M.R., Islam, M.N., Huque, M.M., Hamdan, S., Ahmed, A.S.: Effect of chemical treatment on rice husk (rh) reinforced polyethylene (pe) composites. BioResources **5**, 854–869 (2010)
17. Sinha, E., Rout, S.K.: Influence of fibre-surface treatment on structural, thermal and mechanical properties of jute fibre and its composite. Bull. Mater. Sci. **32**(1), 65–76 (2009). https://doi.org/10.1007/s12034-009-0010-3

18. Thakur, V.K., Thakur, M.K., Gupta, R.K.: Graft copolymers of natural fibres for green composites. Carbohydrates Polymers **104**(1), 87–93 (2014). https://doi.org/10.1016/j.carbpol.2014.01.016
19. Jagtap, S.B., Mohan, M.S., Shukla, P.G.: Improved performance of microcapsules with polymer nanocomposite wall: preparation and characterization. Polymers **83**(1), 27–33 (2016). https://doi.org/10.1016/j.polymer.2015.12.011
20. Islam, M.S., Jeffrey, S.C., Miao, M.: Effect of removing polypropylene fibre surface finishes on mechanical performance of kenaf/polypropylene composites. Compos. Part A: Appl. Sci. Manuf. **42**(11), 1687–1693 (2011). https://doi.org/10.1016/j.compositesa.2011.07.023
21. Bonnia, N.N., Mahad, M.M., Surip, S.N., Anuar, H.: Polyester/kenaf composite; effect of matrix modification. Eng. Ind. Appl. **12**(1), 518–522 (2012). https://doi.org/10.1109/ISBEIA.2012.6422940
22. Azwa, Z.N., Yousif, B.F.: Thermal degradation study of kenaf fibre/epoxy composites using thermo gravimetric analysis. Mater. Polym. Compos. **16**, 256–264 (2013)
23. Anuar, H., Hassan, N.A., Fauzey, M.: Compatibilized PP/EPDM-kenaf fibre composites using melt blending method. Adv. Mater. Res. **1**, 743–747 (2011). https://doi.org/10.4028/www.scientific.net/AMR.264-265.743
24. Lin, J., Yang, Z., Hu, X., Hong, G., Zhang, S., Song, Wei: The effect of alkali treatment on properties of dopamine modification of bamboo fibre/polylactic acid composites. Polymers **10**(4), 403–407 (2018). https://doi.org/10.3390/polym10040403
25. Ray, D., Sengupta, S., Sengupta, S.P., Mohanty, A.K., Misra, M.: A study of the mechanical and fracture behavior of jute fabric reinforced clay modified thermoplastic starch-matrix composites. Macromol. Mater. Eng. **292**(10–11), 1075–1084 (2007). https://doi.org/10.1002/mame.200700111
26. Hossen, M.F., Hamdan, S., Rahman, M.R., Rahman, M.M., Liew, F.K., Lai, J.C.: Effect of fibre treatment and nano-clay on the tensile properties of jute fibre reinforced polyethylene/clay nanocomposites. Fibres Polym. **16**(1), 479–485 (2015). https://doi.org/10.1007/s12221-015-0479-x
27. Kumar, V., Kumar, R.: Improved mechanical and thermal properties of bamboo-epoxy nanocomposites. Polym. Compos. **33**(3), 362–370 (2012). https://doi.org/10.1002/pc.22155
28. Adamu, M., Rahman, M.R., Hamdan, S.: Formulation optimization and characterization of bamboo/polyvinyl alcohol/clay nanocomposite by response surface methodology. Compos. B Eng. **176**(1), 1–9 (2019). https://doi.org/10.1016/j.compositesb.2019.107297

Polylactic Acid Activated Bamboo Carbon Nanocomposites

Md Rezaur Rahman, Sinin Hamdan, and Muhammad Khusairy Bin Bakri

Abstract In this study, activated bamboo carbon was used to prepare nanocomposites. The nanocomposites were characterizing by the Fourier transform infrared spectroscopy (FTIR), x-ray diffraction (XRD), scanning electron microscopy (SEM) and energy dispersive x-ray spectroscopy (EDS). The mechanical test was carried out in this study. The 2202.71 to 3078.39 cm^{-1} band represent the whole activated carbon functional group. The smooth surface morphology indicates that the PLA is compatible with the activated carbon and less agglomeration to the structure. Bamboo carbon nanocomposites showed higher Young's modulus compare to raw bamboo itself.

Keywords Bamboo · Carbon · Poly Lactic Acid · Nanocomposites · Optimization

1 Introduction

Emerging plastic and ceramics materials market vastly plays a big role nowadays in the world. The world plastic industries growth is said to be steady. It also gains lots of attentions in the industrial or the household applications, whereas most organization used it to start new business, which able to penetrate the market with their characteristics and own criteria [1]. Cost-effectiveness material that consumer needed are usually weight saving and satisfy in every property, as nowadays people want things, which are valuable and long last used [1]. Polylactic acid (PLA) bioplastics is considered as biodegradable and are used as a packaging due its physical and chemical property, which is suitable to be used in many applications [2–5]. In biomedical, PLA may have potential to be used with a living tissue, whereas it can be implanted it to the skin of the patient, as it is compatible and not harmful [6, 7]. PLA advantage was its degradation does not involve many removal processes, which is simply done using hydrolysis [8]. The PLA had simplicity that can be obtained or reacted with the renewable sources. Researchers are more interested to start or continuing

M. R. Rahman (✉) · S. Hamdan · M. K. B. Bakri
Faculty of Engineering, Universiti Malaysia Sarawak, Jalan Datuk Mohammad Musa, 94300, Kota Samarahan, Sarawak, Malaysia
e-mail: rmrezaur@unimas.my

M. R. Rahman (ed.), *Bamboo Polymer Nanocomposites*, Engineering Materials,
https://doi.org/10.1007/978-3-030-68090-9_4

research based on PLA and its importance, especially in terms of packaging, whereas it has higher tensile strength and modulus properties, which affected the conventional polymer system in produces plastic and their respective applications [9]. Most of the PLA characteristic discovered was more towards of introduced different filler into the polymer.

For instance, in China, most of the bamboo's available are used to produce households, composite, boards, floors, furniture, and other products as a replacement for wood based [10]. Nowadays, many bamboo companies continued to support other countries activities such as marketing, harvesting and manufacturing, especially Bangladesh's, throughout their economic development [11]. According to Marsh and Smith [12] the advent demand of bamboo used as replacement for wood material added a balanced rising demand for its competitor, wood. Furthermore, compared with wood, bamboo is relatively cheaper. As a result, Harmonized Scheme has been introduced and bamboo-based products have been listed in trade codes, especially those with timber [13]. The modern latest development of bamboo composite products is linked to some studies through the bamboo optimization functionally graded material (FGMs), which are used in advance composites production [14].

In contrast, based on Mohanty et al. [15], bamboo is classified as grass fiber/bio fiber and has potential to be used in automotive composite applications. Therefore, in advanced industries, a raw bamboo material absolutely contributed and highly utilized. Widely worldwide, significant number of researches have studied the activated carbon (ACs) synthesis, modification, characterization and application, which were obtained from lignocellulosic resources, i.e. wood and coconut shells [16]. More than 300,000 t/year and high yielded world production, the most common precursors for the large scale of synthesis was AC [17]. As low-cost material, the development of the AC in terms of the selection of the precursors depends upon many factors. To fitted and become the most preferably lignocellulosic materials, it must be freely available, inexpensive and non-hazardous in nature, whereas in addition must retain its high content of carbon and low amounts of inorganic materials [18, 19]. Nowadays, the preparation of lignocellulosic origin from AC from the raw materials are increasing worldwide. Several precursors have been selected such as nuts and almond shells, bamboo, rice straw and husk, apricot stones, wheat bran, oil palm tree, etc. [18, 19].

As stated, bamboo has advantages used as a precursor to the AC and benefit the process and its products. One of the bamboo advantages as the AC precursors was relatively due to its fast-growing rate, since it provides a renewable source of raw material. Bamboo are characterized as grass family group of woody perennial plants of *poaceae*, which is a subfamily of *bambusoideae* that includes more than 1200 species and more than 100 genders [20, 21]. Therefore, this chapter discussed the results of the activated bamboo carbon polylactic nanocomposites through characterization, i.e. physico-chemical morphological analysis, and mechanical properties. The degree of changes based on the result are shown in the modification process.

2 Methodology

2.1 *Materials*

Bamboo, nitrogen gas, and distilled water was prepared and obtained from Universiti Malaysia Sarawak, Kota Samarahan, Sarawak, Malaysia. Potassium hydroxide (KOH) and polylactic acid (PLA) was obtained from Sigma Aldrich (US) Company.

2.2 *Preparation of Bamboo, Activated Carbon and Its Composites*

The bamboo was cut into short strip and dried in the oven at 45°C to 70°C for 1 to 2 weeks. The short small dried bamboo strip was ground into small sizes and carbonized at 400°C for 2 h under no air flow conditions. The bamboo was then crushed and sieved into powdered size. The carbonized bamboo was activated using 25% by volume of KOH solution and the weight ratio of the carbonized bamboo to KOH is 1:3. The mixture was dehydrated at 200°C until became slurry. Then, the mixture was activated in a tube-type electronic heating furnace under nitrogen gas with flow rate at 200 mL/min. It was heated up to 400°C using a heating rate of 10°C/min. During the activation, the temperature was maintained at 400°C for 1 h. After the activation, the product is wash using distilled water several times and dried again in the oven. The activated carbon was then mixed with PLA at different percentage and were hot press using hot press machine at 140°C. It was cooled down with cold pressed at 27°C for curing.

2.3 *Tensile Test*

The tensile mechanical properties of the PLA/activated bamboo carbon nanocomposites were measured and plotted. The tests were done using Shimadzu MSC-5/500 universal testing machine (Kyoto, Japan) operating at a crosshead speed of 5 mm/min according to ASTM D638-14 [22] standards.

2.4 *Fourier Transform Infrared Spectroscopy (FTIR)*

The PLA/activated bamboo carbon nanocomposites were analyzed using a Fourier transform Infrared Spectrometer (FTIR) within the wavenumber ranging from 4000 to 400 cm^{-1} to identify the functional group presence in the sample composition. The tests were done according to ASTM E168-16 [23] and ASTM E1252-98 [24].

2.5 *Scanning Electron Microscopy (SEM) and Energy Dispersive X-Ray Spectroscopy (EDX/EDS)*

The morphological properties of the PLA/activated bamboo carbon nanocomposites were obtained from the scanning electron microscopy (SEM) machine to evaluate the surface morphology and element from the samples. Both tests were conducted according to ASTM E2015-04 [11] and ASTM E1508-12 [12] standards. The EDS was repeated multiple times to get the most repeated results.

3 Results and Discussions

3.1 *FTIR Analysis*

Figures 1, 2, 3, 4, 5, 6, and 7 shows the PLA/Activated bamboo nanocomposites at different weigh, while Figs. 8 and 9 shows the AC and PLA, respectively. In Figs. 1–7, the peak at 1523.76 cm^{-1} shows that there was a strong vibration of N-O bond stretching which appear to be the nitro compound. There are two functional existed on the 1695.43 cm^{-1} peak band, which shows strong C = O vibration that suggest the strong interaction of the activated carbon [18, 19]. Even though C-H function group also present in the compound, a weak vibration suggested that aromatic compound was started to occur at this peak band, which indicate the polylactic acid properties is started to appear. The peak 2160.27 cm^{-1} band indicated a strong vibration of azide

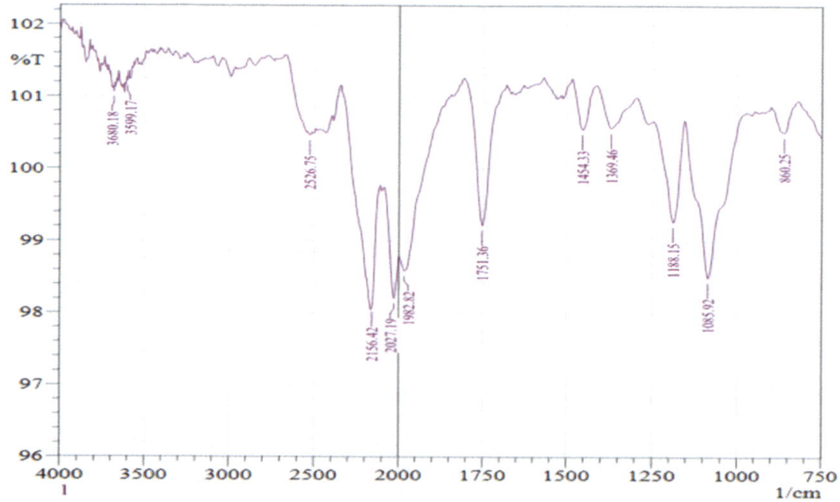

Fig. 1 Sample 1 (3.33 g AC and 6.67 g PLA)

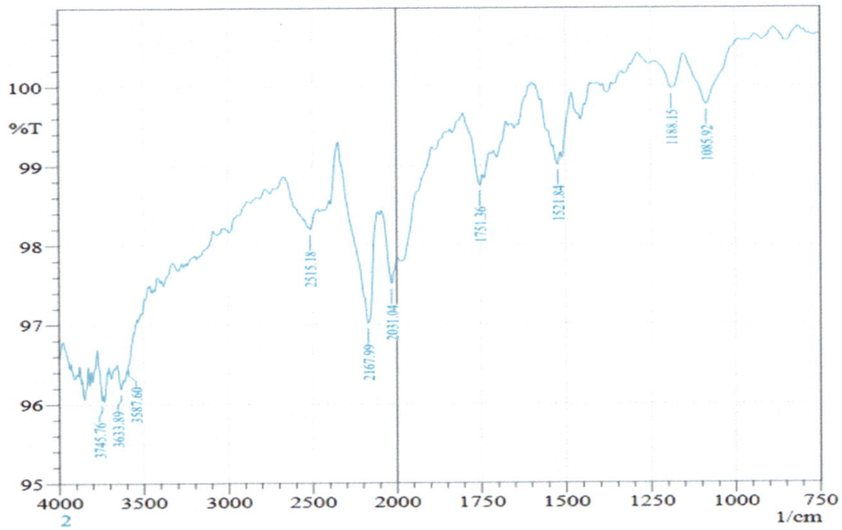

Fig. 2 Sample 2 (3 g AC & 7 g PLA)

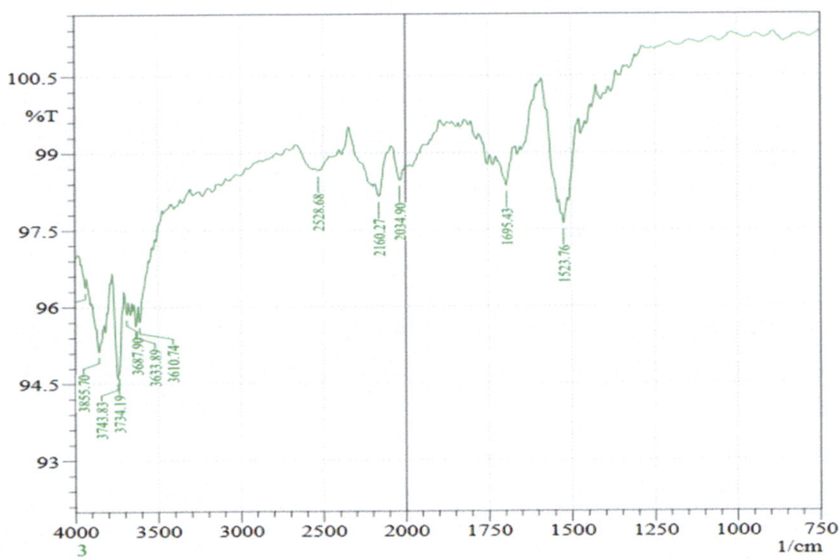

Fig. 3 Sample 3 (5 g AC & 5 g PLA)

compound, while the 2528.68 cm^{-1} peak band shows weaker vibration that suggested the appearance of the thiol, which was the S-H functional group. The medium and sharp 3610.74 to 3938.64 cm^{-1} band was due to O-H vibration functional group, which indicated the presence of alcohol compound for polylactic acid.

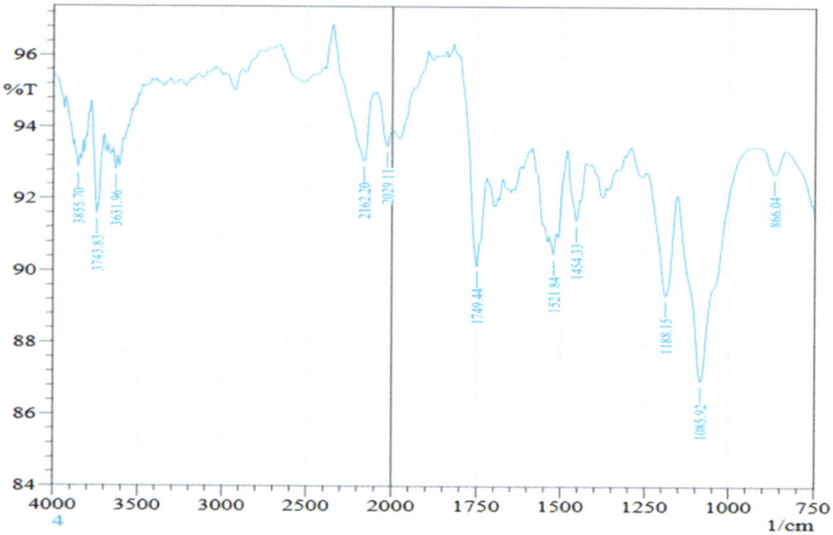

Fig. 4 Sample 4 (6 g AC & 4 g PLA)

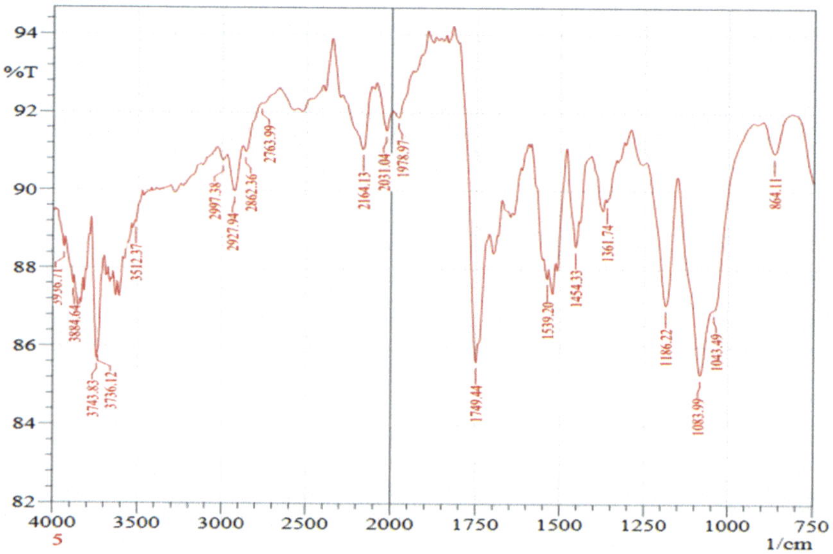

Fig. 5 Sample 5 (4 g AC & 6 g PLA)

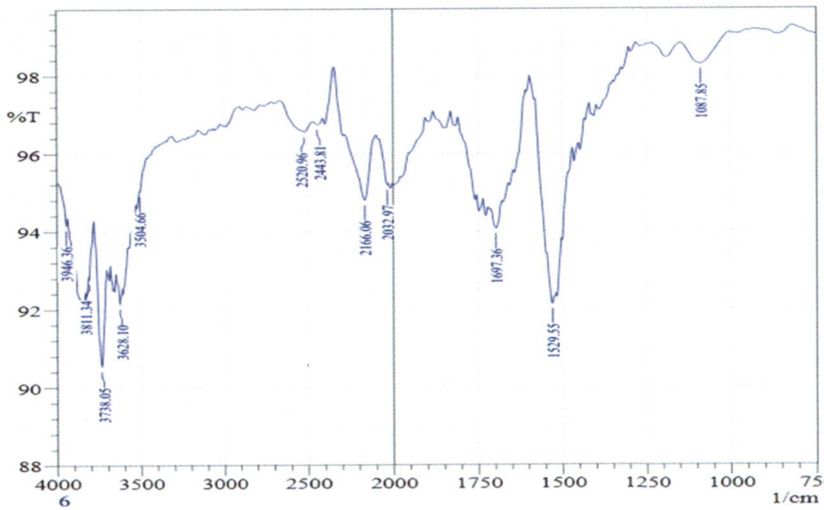

Fig. 6 Sample 6 (4.67 g AC & 5.33 g PLA)

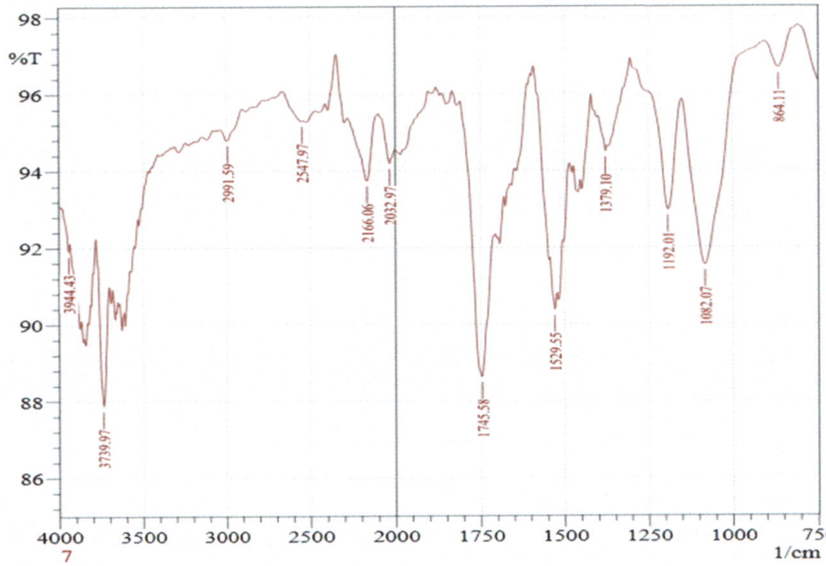

Fig. 7 Sample 7 (2 g AC & 8 g PLA)

In Fig. 8, the 1097.50 cm^{-1} peak band spectra show the strong presence of the C-O vibrations and stretching functional group of the secondary alcohol, as the KOH solution start to saturate to the bamboo. On the medium peak band, there was also C-N vibration and stretching functional group, which indicated the presence of

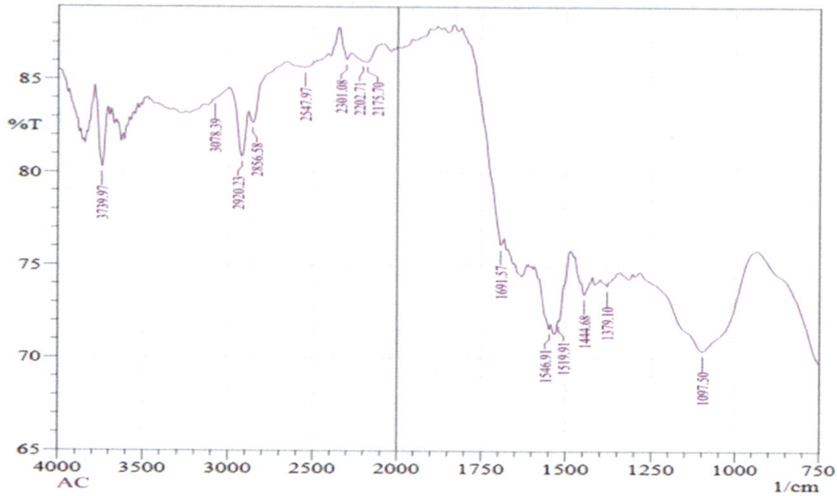

Fig. 8 Sample of Activated Carbon (AC)

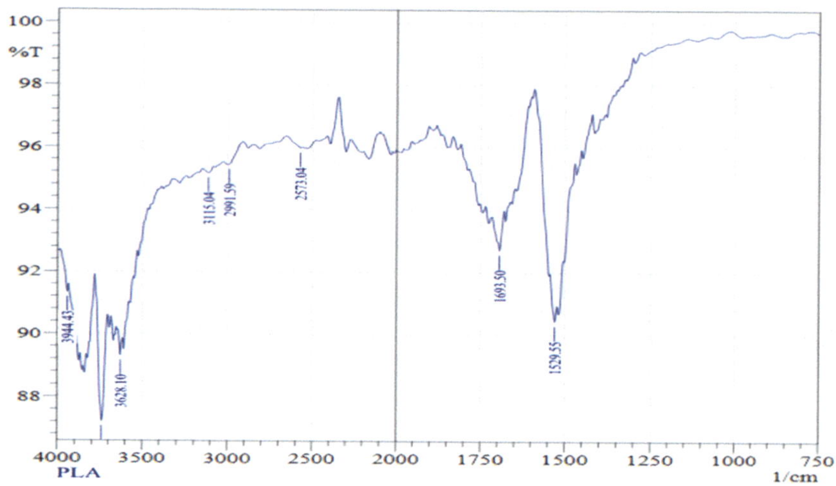

Fig. 9 Sample of Polylactic Acid (PLA)

amine compound. The 1379.10 cm^{-1} peak band shows two functional group that were present along the spectrum of the activated carbon. It was O-H with medium vibration functional group and phenol molecule group, which was from the activating agent such, i.e. NaOH and KOH. The strong $S = O$ stretching vibration functional group also detected, which identify the presence of the sulfonyl chloride. The 1519.91 to 1546.91 cm^{-1} band shows strong nitro vibration and stretching compound, whereas the molecule was turned into other form [18, 19]. The 2202.71 to 3078.39 cm^{-1} band represent the whole activated carbon functional group. The C-H functional group and the medium intermolecular suggested the appearance of the alkane, alkene and C $\equiv$ C stretching at 2202.71 cm^{-1} band, but very weak in molecule vibration which indicated the presence of the alkyne.

Figure 9 shows the peak of polylactic acid (PLA) at 1529.55 cm^{-1} band, which had a strong N-O stretching of the functional group, which consist of the nitro compound. The 1693.50 cm^{-1} peak band shows C = O stretching functional group, which indicate the conjugated aldehyde compound with strong vibration of the molecule. The 2573.04 cm^{-1} peak band shows the S-H weak vibration functional and stretching group which indicated the acetylation. While the 2991.59 cm^{-1} peak band shows the alkane compound existed with medium vibration of stretching. the 3628.10 to 3944.43 cm^{-1} band shows a medium O-H stretching vibration functional group, which provided the characteristics of the alcohol compound in the mixture [18, 19].

3.2 *Scanning Electron Microscopy (SEM)*

Figure 10a–f show the SEM images of the sample surface, which was rough because of the activated carbon interacted with the polylactic acid. This also indicated better interfacial bonding between the surface of the activated carbon and the polymer matrix in the system of the composite. Different composition provides different surface morphology but the same type of interaction between the fillers, which was activated carbon made from bamboo. This also show better filler dispersion in the composites. Figure 10g shows clearer and smoother surface because of the difference composition, whereas the composition in Fig. 10g were greater in amount of the PLA, which provides the smooth surface. It also noted that the activated carbon as a precursor had a great type of attachment of the molecule into it considering the bamboo are form the natural resources. The smooth surface indicates that the PLA is attached to the macropores of the activated carbon and less agglomeration to the structure.

Figure 11 shows the activated carbon clearly had a rough surface, due to many voids and lots of macropores, which let the particle or substance to attached to the surface of the activated carbon and filled into the AC. AC surface area were larger and the volume of the total pores is high in every activated carbon structure. Figure 12 shows the SEM image for the PLA, which the surface was smooth and had lesser agglomeration. PLA also have more dispersed molecule in the matrix.

3.3 *Energy Dispersive X-Ray Spectroscopy (EDS) Analysis*

From Fig. 13 and Table 1, the mixture composition mainly contained carbon, oxygen, nickel, potassium, and phosphorus. The carbon and oxygen were the main elements in the activated carbon and the polylactic acid. Nickel and phosphorus elements were due to the mix of the activated carbon with the polylactic acid. While potassium was

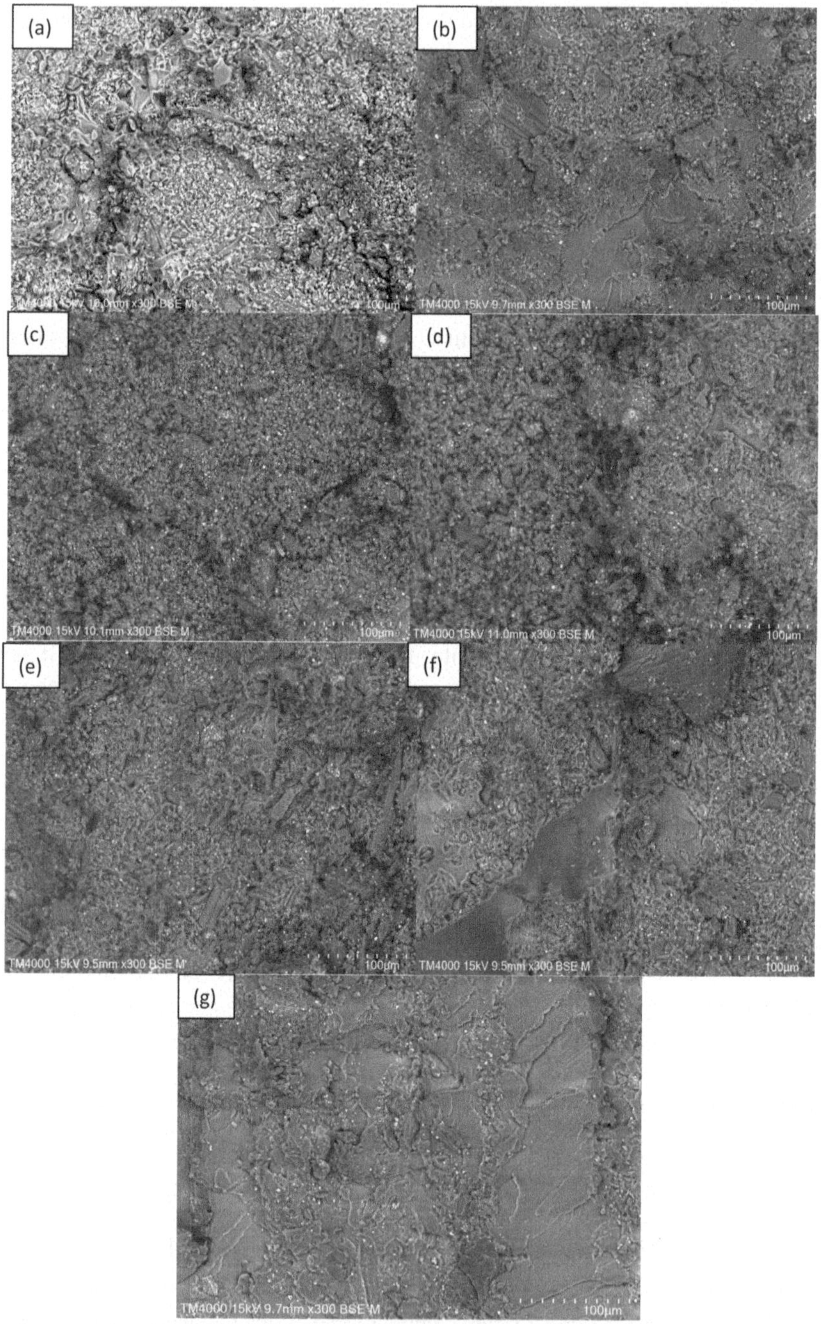

Fig. 10 SEM images of **a** sample 1, **b** sample 2, **c** sample 3, **d** sample 4, **e** sample 5, **f** sample 6, and **g** sample 7

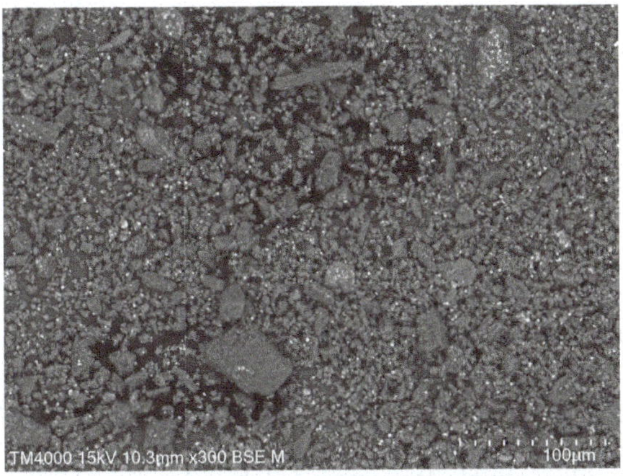

Fig. 11 300x magnification of activated carbon (AC)

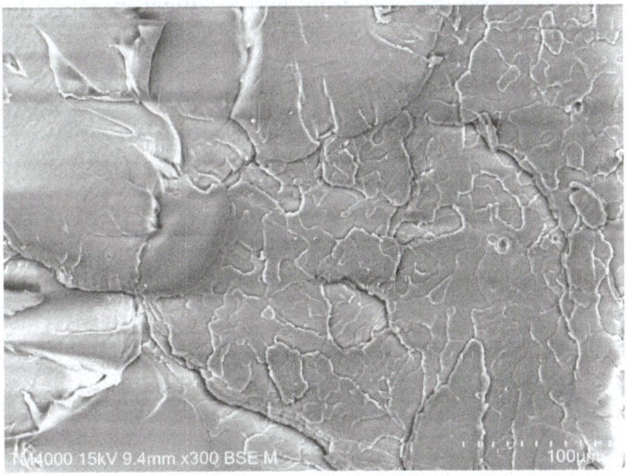

Fig. 12 300x magnification of polylactic acid (PLA)

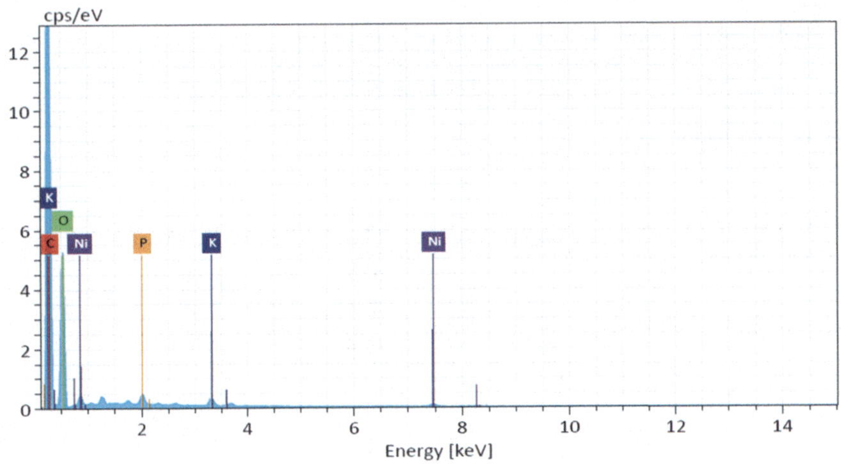

Fig. 13 EDS for sample 1

Table 1 Element and mass of sample 1

Element	Mass (%)	Atom (%)
C	60.40	67.91
O	37.12	31.33
Ni	1.24	0.28
K	0.74	0.26
P	0.50	0.22

from the activating agent for the activated carbon, i.e. KOH. From the Fig. 14 and Table 2, to Fig. 19 and Table 7 the same mixture compound element was shown, which is carbon, oxygen, nickel, phosphorus, calcium, and magnesium (Figs. 15, 16, 17, and 18). Carbon and oxygen are the main element, which appear in every EDS result of sample 2 to sample 7 (Tables 3, 4, 5, and 6). The other substitute which is the nickel, phosphorus, calcium, and magnesium were the production of the mixture of

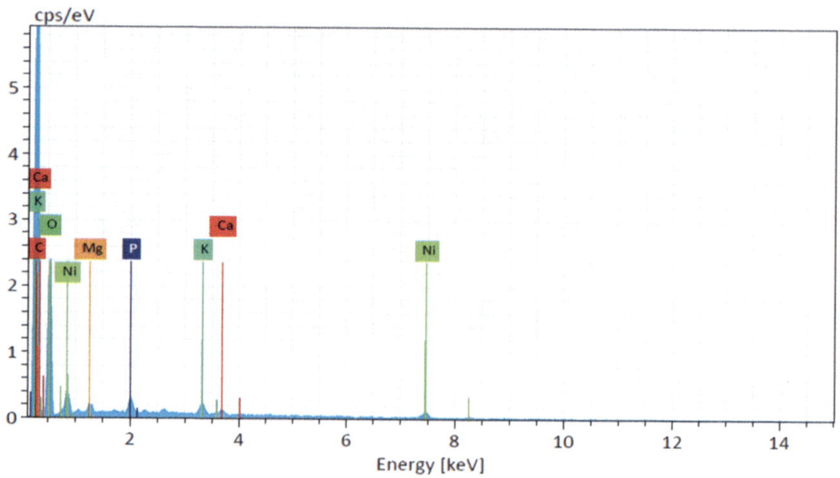

Fig. 14 EDS for sample 2

Table 2 Element and mass of sample 2

Element	Mass (%)	Atom (%)
C	58.37	67.07
O	35.91	30.97
Ni	2.47	0.58
K	1.08	0.38
P	1.02	0.46
Mg	0.66	0.38
Ca	0.48	0.17

the two compounds of between activated carbon and polylactic acid. In the process, the carbon attachment and whole structure became the final product, which was due to the activated agent, which were reacted with the functional groups, such as the phosphorus, hydroxyl and carboxyl that formed in the activated carbon active surface. This also indicated the presence of the element such as the phosphorus and nickel, which was due to the preparation of the activated carbon, whereas the reaction of the catalytic process. Figure 20 and Table 8 shows that the activated carbon elements were the main element existed in the activated carbon from bamboo. It also mainly contains oxygen, potassium, magnesium, phosphorus, and calcium. The PLA was obtained from a wheat starch, whereas PLA mainly made from plant derivatives, which contain the element carbon, oxygen, and nickel, as shown in Fig. 21 and Table 9.

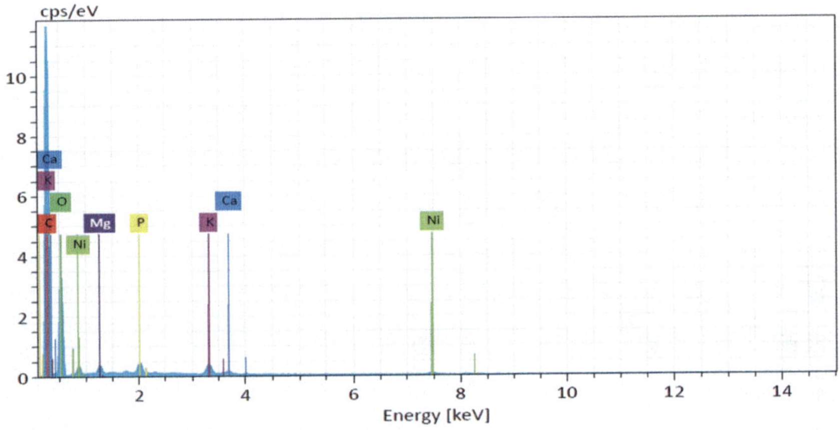

Fig. 15 EDS for sample 3

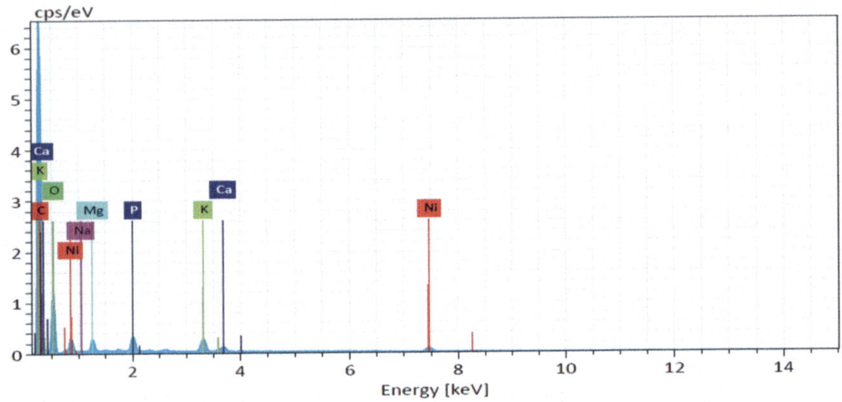

Fig. 16 EDS for sample 4

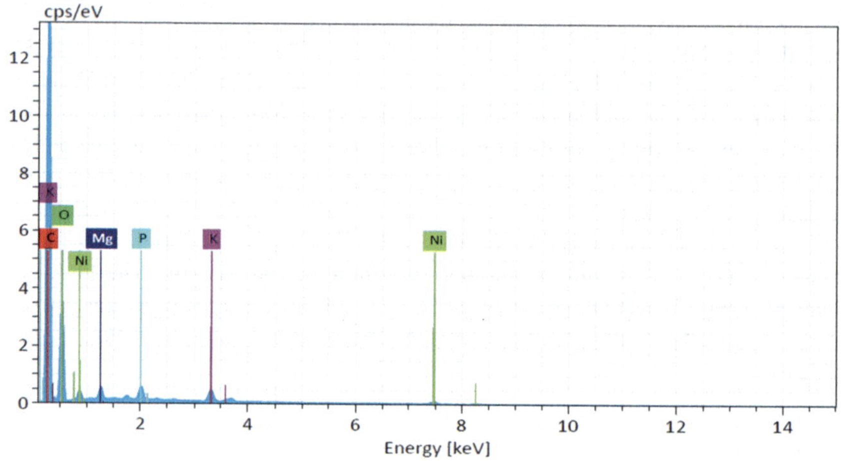

Fig. 17 EDS for sample 5

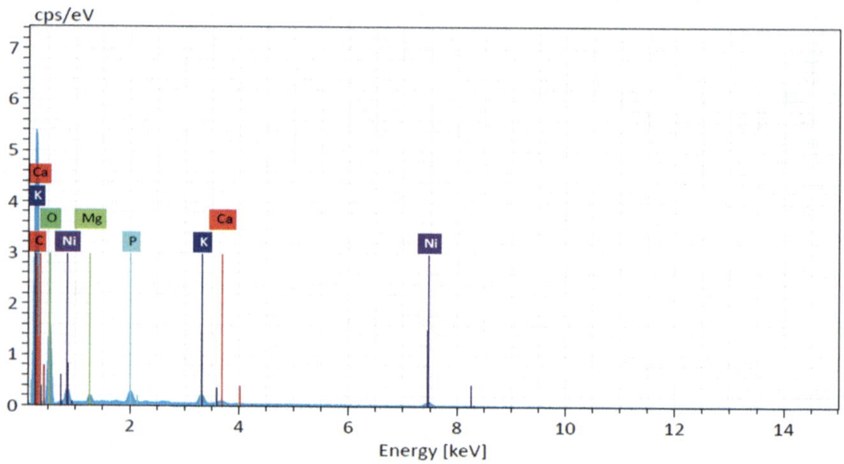

Fig. 18 EDS for sample 6

Table 3 Element and mass for sample 3

Element	Mass (%)	Atom (%)
C	62.31	70.20
O	33.46	28.30
Ni	1.24	0.29
K	1.24	0.43
P	0.70	0.31
Mg	0.60	0.33
Ca	0.45	0.15

Table 4 Element and mass for sample 4

Element	Mass (%)	Atom (%)
C	67.01	75.74
O	25.38	21.54
Ni	2.42	0.56
K	1.96	0.68
P	1.30	0.57
Mg	1.11	0.62
Ca	0.78	0.26

Table 5 Element and mass for sample 5

Element	Mass (%)	Atom (%)
C	65.17	72.29
O	30.72	25.76
Ni	1.46	0.33
K	1.27	0.43
P	0.71	0.31
Mg	0.68	0.38

Table 6 Element and mass for sample 6

Element	Mass (%)	Atom (%)
C	61.39	70.27
O	32.00	27.50
Ni	2.64	0.62
K	1.68	0.59
P	1.05	0.47
Ca	0.67	0.23
Mg	0.57	0.32

3.4 Tensile Test

From Figs. 22 to 23 shows the Young modulus and tensile strength for each sample. For Sample 1 that 3.33 g of activated carbon and 6.67 g of polylactic acid had the highest value of the young modulus. While, Sample 7 obtained the highest tensile strength. Sample 6 had the lowest tensile strength and Young modulus value. The highest tensile strength was from the sample 7 with 11.98 MPa, which meaning that the strength of the molecule is strongest, and the molecule takes longer time to fracture when applied pressured [20].

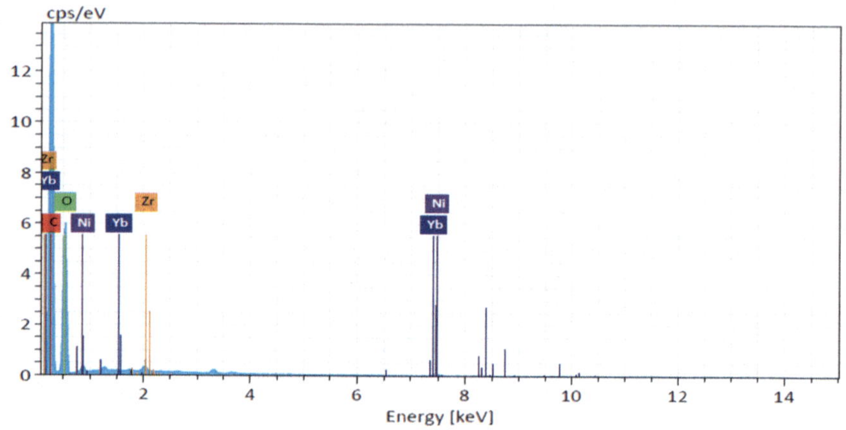

Fig. 19 EDS for sample 7

Table 7 Element and mass for sample 7

Element	Mass (%)	Atom (%)
C	58.91	66.29
O	39.55	33.41
Ni	0.96	0.22
Zr	0.47	0.07
Yb	0.10	0.01

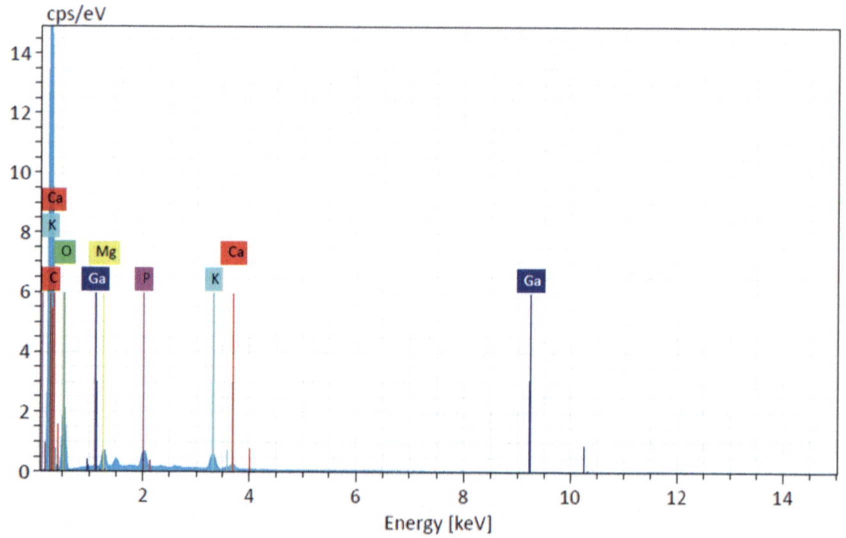

Fig. 20 EDS for Activated Carbon (AC)

Table 8 Element and mass for activated carbon

Element	Mass (%)	Atom (%)
C	73.00	79.65
O	22.80	18.68
K	1.60	0.54
Mg	0.96	0.52
P	0.94	0.40
Ca	0.65	0.21

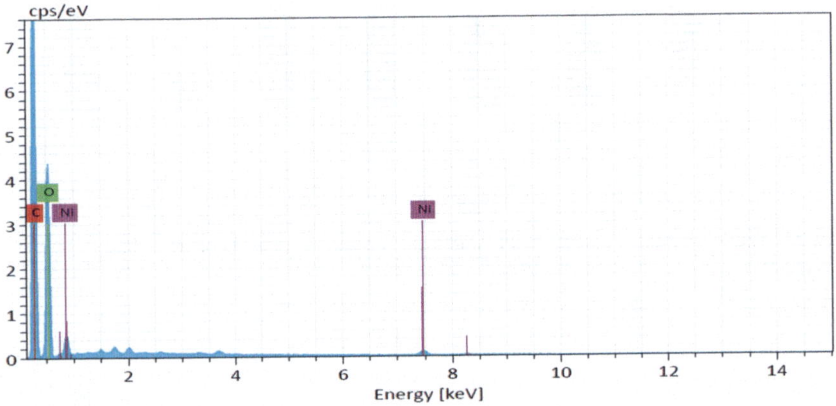

Fig. 21 EDS for Polylactic Acid (PLA)

Table 9 Element and mass for polylactic acid (PLA)

Element	Mass (%)	Atom (%)
C	55.12	63.10
O	42.21	36.28
Ni	2.67	0.63

4 Conclusions

In conclusion, polylactic acid activated bamboo carbon nanocomposites showed improved properties compare to raw bamboo. FT-IR spectra showed the carbonyl vibrations and stretching functional group of the secondary alcohol, as the KOH solution start to saturate to the bamboo whereas activated bamboo carbon nanocomposites showed highest mechanical properties. The smoother surface was observed in activated bamboo carbon nanocomposites compare to others sample.

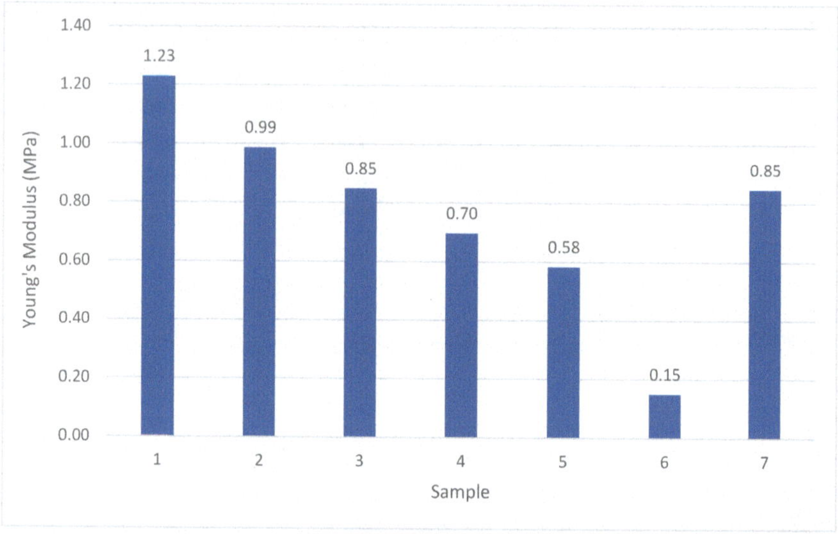

Fig. 22 Young's modulus sample for PLA/activated bamboo carbon nanocomposites

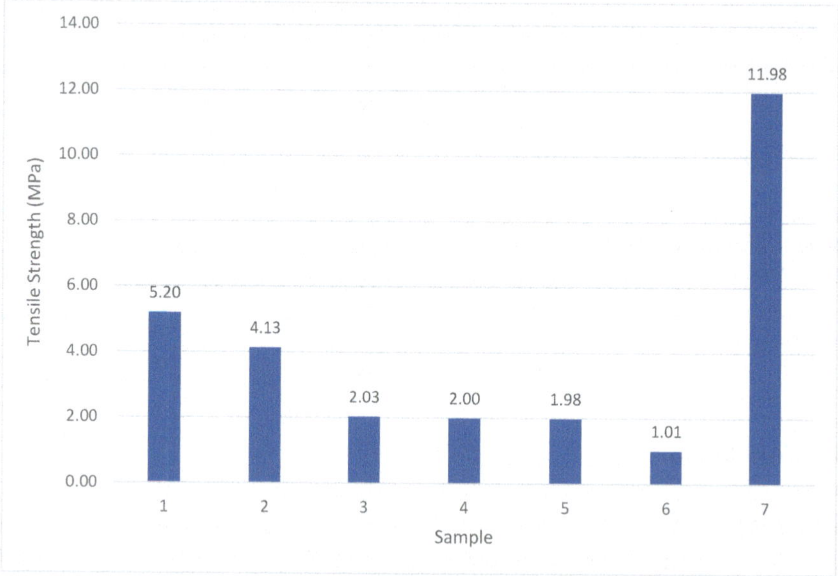

Fig. 23 Tensile strength sample for PLA/activated bamboo carbon nanocomposites

Acknowledgements The authors would like to acknowledge Universiti Malaysia Sarawak (UNIMAS) for the support.

References

1. Rodrigues, C.S.D., Soares, O.S.G.P., Pinho, M.T., Pereira, M.F.R., Madeira, L.M.: P-nitrophenol degradation by heterogeneous fenton's oxidation over activated carbon-based catalysts. Appl. Catal. B: Environ. **219**(1), 109–122 (2017). https://doi.org/10.1016/j.apcatb.2017.07.045
2. Jem, J.K., Tan, B.: The development and challenges of poly(lactic acid) and poly(glycolic acid). Adv. Ind. Eng. Polym. Res. **3**(2), 60–70 (2020). https://doi.org/10.1016/j.aiepr.2020.01.002
3. Jimenez, L., Mena, M.J., Prendiz, J., Salas, L., Vega-Baudrit, J.: Polylactic acid (PLA) as a bioplastic and its possible applications in the food industry. J. Food Sci. & Nutr. **5**(1), 48 (2019). https://doi.org/10.24966/FSN-1076/100048
4. Jamshidian, M., Tehrany, E.A., Imran, M., Jacquot, M., Desobry, S.: Poly-lactic acid: production, applications, nanocomposites, and release studies. Compr. Rev. Food Sci. Food Saf. **9**(5), 552–571 (2010). https://doi.org/10.1111/j.1541-4337.2010.00126.x
5. Reichert, C.L., Bugnicourt, E., Coltelli, M.-B., Cinelli, P., Lasseri, A., Canesi, I., Braca, F., Martinez, B.M., Alonso, R., Agostinis, L., Verstichel, S., Six, L., Mets, S.D., Gomez, E.C., Ißbrücke, C., Geerinck, R., Nettleton, D.E., Campos, I., Sauter, E., Pieczyck, P., Schmid, M.: Bio-based packaging: materials, modifications, industrial applications and sustainability. Polymers **12**(7), 1558 (2020). https://doi.org/10.3390/polym12071558
6. Song, R., Murphy, M., Li, C., Ting, K., Soo, C., Zheng, Z.: Current development of biodegradable polymeric materials for biomedical applications. Drug Des., Dev. Ther. **12**(1), 3117–3145 (2018). https://doi.org/10.2147/DDDT.S165440
7. Sha, L., Chen, Z., Chen, Z., Zhang, A., Yang, Z.: Polylactic acid based nanocomposites: promising safe and biodegradable materials in biomedical field. Int. J. Polym. Sci. **1**, 1–11 (2016). https://doi.org/10.1155/2016/6869154
8. Auras, R., Harte, B., Selke, S.: An overview of polylactides as packaging materials. Macromol. Biosci. **4**(9), 835–864 (2004). https://doi.org/10.1002/mabi.200400043
9. Sodergard, A., Stolt, M.: Properties of lactic acid-based polymers and their correlation with the different fillers. Prog. Polym. Sci. **27**(6), 1123–1163 (2002). https://doi.org/10.1016/S0079-6700(02)00012-6
10. Chaowana, P.: Bamboo: an alternative raw material for wood and wood-based composites. J. Mater. Sci. Res. **2**(2), 90–102 (2013). https://doi.org/10.5539/jmsr.v2n2p90
11. Rana, M.P., Mukul, S.A., Sohel, M.S.I., Shaheed, M., Chowdhury, H., Akhter, S., Chowdhury, M.Q., Koike, M.: Economics and employment generation of bamboo-based enterprises: a case study from Eastern Bangladesh. Small-Scale For. **9**(1), 41–51 (2009). https://doi.org/10.1007/s11842-009-9100-8
12. Marsh, J., Smith, N.: New bamboo industries and pro-poor impacts: lesson from china and potential for Mekong countries. A Cut for the Poor, Proceeding of the International Conference on Managing Forests for Poverty Reduction: Capturing Opportunities in Forest Harvesting and Wood Processing for the Benefit of the Poor. Ho Chi Minh City, Vietnam, 216–232 (2007). https://ccsindia.org/bamboo/4o-new_bamboo_industries_and_pro-poor_impact-china.pdf
13. Buckingham, K., Wu, L., Lou, Y.: Can't see the (bamboo) forest for the trees: examining bamboo's fit within international forestry institutions. Ambio **43**(1), 770–778 (2014). https://doi.org/10.1007/s13280-013-0466-7
14. Nogata, F., Takahashi, H.: Intelligent functionally graded material: bamboo. Compos. Eng. **5**(7), 743–751 (1995). https://doi.org/10.1016/0961-9526(95)00037-N

15. Mohanty, A.K., Misra, M., Drzal, L.T.: Sustainable bio-composites from renewable resources: opportunities and challenges in the green materials world. J. Polym. Environ. **10**(1), 19–26 (2002). https://doi.org/10.1023/A:1021013921916
16. Nor, N.M., Lau, L.C., Lee, K.T., Mohamed, A.R.: Synthesis of activated carbon from lignocellulosic biomass and its applications in air pollution control—a review. J. Environ. Chem. Eng. **1**(4), 658–666 (2013). https://doi.org/10.1016/j.jece.2013.09.017
17. Maurao, P.A.M., Laginhas, C., Custodio, F., Nabais, J.M.V., Carrott, P.J.M., Ribeiro Carrott, M.M.L.: Influence of oxidation process on the adsorption capacity of activated carbons from lignocellulosic precursors. Fuel Process. Technol. **92**(2), 241–246 (2011). https://doi.org/10.1016/j.fuproc.2010.04.013
18. Hayashi, J., Kazehaya, A., Muroyama, K., Watkinson, A.P.: Preparation of activated carbon from lignin by chemical activation. Carbon **38**(13), 1873–1878 (2000). https://doi.org/10.1016/S0008-6223(00)00027-0
19. Olivares-Marin, M., Fernandez-Gonzalez, C., Macias-Garcia, A., Gomez-Serrano, V.: Preparation of activated carbons from cherry stones by activation with potassium hydroxide. Appl. Surf. Sci. **252**(17), 5980–5983 (2006). https://doi.org/10.1016/j.apsusc.2005.11.018
20. Gonzalez, P.G., Hernandez-Quiroz, T., Garcia-Gonzalez, L.: The use of experimental design and response surface methodologies for the synthesis of chemically activated carbons produced from bamboo. Fuel Process. Technol. **127**(1), 133–139 (2014). https://doi.org/10.1016/j.fuproc.2014.05.035
21. Krzesinska, M., Pilawa, B., Pusz, S., Ng, J.: Physical characteristics of carbon materials derived from pyrolysed vascular plants. Biomass Bionergy **30**(2), 166–176 (2006). https://doi.org/10.1016/j.biombioe.2005.11.009
22. ASTM D638-14: Standard test method for tensile properties of plastics, ASTM International, West Conshohocken, PA (2014). https://doi.org/10.1520/D0638-14
23. ASTM E168-16: Standard practices for general techniques of infrared quantitative analysis, ASTM International, West Conshohocken, PA (2016). https://doi.org/10.1520/E0168-16
24. ASTM E1252-98: Standard practices for general techniques for obtaining infrared spectra for qualitative analysis, ASTM International, West Conshohocken, PA (2013).

Investigation on the Brittle and Ductile Behavior of Bamboo Nano Fiber Reinforced Polypropylene Nanocomposites

Md Rezaur Rahman, Sinin Hamdan, and Muhammad Khusairy Bin Bakri

Abstract Research was carried out focuses on the change of bamboo fiber behavior, from brittle to ductile, with the help of thermoplastics polypropylene. Bamboo nanofiber was extracted from raw bamboo and underwent 5% of sodium hydroxide (NaOH) chemical treatment for delignification process. Next, through hot pressed method, the bamboo nanofiber was reinforced with polypropylene, whereas eight different bamboo nanofiber composition reinforced polypropylene were prepared. The samples were analyzed using tensile test, scanning electron microscopy (SEM), energy dispersive X-ray analysis (EDX), Fourier transform spectroscopy (FTIR), and optimization model by using Design Expert software. An improvement results was shown in the tensile strength, range from 10.27 to 80.12%. While for the bamboo nanofiber reinforced ductility, the calculation of elongation at breaks showed increases from 3 to 42%. Based on the experiment, the best sample was Sample 3, which consist of 93.73% of polypropylene, 3.28% of nanofiber, and 3% of MMT as the filler. It gave the highest tensile strength of 18.9914 MPa and second highest ductility of 181.57%.

Keywords Bamboo · MMT · Polypropylene · Nanocomposites · Optimization

1 Introduction

As a sustainable fiber, bamboo fiber has gained a significant amount of interest [1–3]. It was known as a cellulosic fiber, as it was extracted from the plant of natural bamboo [1–3]. After cotton, silk, linen and wool, bamboo was the fifth natural green fibers, which have an exceptional biodegradable material. It was also comparable in strength with the conventional glass fibers [4–6]. According to Okubo et al. [7], among the researchers, bamboo fiber was often called as the natural glass fiber due to its properties. Furthermore, the bamboo fiber preparation usually required 3 to 4 years old bamboo, which considered mature [8]. Most of the bamboo produced

M. R. Rahman (✉) · S. Hamdan · M. K. B. Bakri
Faculty of Engineering, Universiti Malaysia Sarawak, Jalan Datuk Mohammad Musa, 94300, Kota Samarahan, Sarawak, Malaysia
e-mail: rmrezaur@unimas.my

M. R. Rahman (ed.), *Bamboo Polymer Nanocomposites*, Engineering Materials,
https://doi.org/10.1007/978-3-030-68090-9_5

during the whole process underwent alkaline hydrolysis, multiphase bleaching of bamboo stems and leaves, as well as a chemical treatment of the starchy pulp that are generated [8]. The advantage of bamboo fiber was being elastic, environmental-friendly, biodegradable, higher moisture absorption, and having various of micro gaps, which softer than cotton [8, 9]. In addition, bamboo has a substantial tensile strength, which were highly durable, tough and stable, and this makes it versatile to be applied into various industries, i.e. textile, building structure, and medical apparatus [10–12].

The natural plant fibers reinforced polymer composites application was widely used, especially due to it benefits. Polypropylene (PP) was known with one of the most widely used polymers and was considered as strong as steel, due to its versatility in order to suit the purpose, which can be modified into various ways of the application [13]. It had a linear hydrocarbon structure expressed as C_NH_{2N}. Polypropylene was also known as polypropene, which thermoplastic in nature. It had a tough characteristic, as its mechanically rugged and a resistant to either various base or acid chemical solvents [14, 15]. In many different processing technologies, i.e. injection molding, blow molding, sheet extrusion and thermoforming, the characteristic of PP contributed to the use of it.

The bamboo fiber reinforced with polypropylene brought a lot of advantages. It helps initially to increase the raw bamboo tensile strength from around 15 to 20% increment [16–18]. Nevertheless, the limitation for reinforced composites was seen on high temperature, whereas a crack formation on the materials could happen [19]. This happened with temperature during the manufacturing due to the chain degradation [19]. Hence, the aim of this experiment was to change the bamboo brittle to ductile properties. The method and preparation of the bamboo nanofiber from the raw bamboo as well as the preparation of bamboo nanofiber with polymer, which is polypropylene was also carried out. Therefore, this chapter discussed the results of the bamboo nanofiber polypropylene nanocomposites through characterization, i.e. physico-chemical morphological analysis, and mechanical properties. The degree of changes based on the result are shown in the modification process, which were optimized based on Sodium hydroxide, ethanol, and MMT.

2 Methodology

2.1 Material

The raw bamboo was obtained from the local village located at Kampung Baru, Kota Samarahan, Sarawak, Malaysia. Analytical grade chemicals of polypropylene (PP) and Montmorillonite (MMT) (99% purity, Sigma-Aldrich, St. Louis, MO, USA). Sodium hydroxide (Merck Schuchardt, Hohenbrunn, Germany), and ethanol (95% purity, Braun HmbG, Kronberg, Germany) were also used.

2.2 Extraction of Bamboo Nanofiber from Raw Bamboo

The nanofiber is extracted from the raw bamboo by removed the node of the raw bamboo using saw machine. The remaining part of the internodes were sliced in longitudinal direction into a thin strip with 20 to 30 cm of length, and 2 to 3 mm in thickness, by using slicer. The thin strips were immersed in 5% NaOH solution at 70°C for 10 h. The alkaline treated strips were extracted into fiber bundles, with an average diameter of 200 μm by using the roller looser. Next, to neutralize the fiber bundles, it was washed with fresh water and dispersed in water with a content of 10wt% of ethanol. After that, the mixer was used to cut the fiber bundle into pulp fiber, with an average diameter of 20 μm and aspect ratio of 65. The pulp fiber was then ground 15 times between static grind and rotating grind of 1500 rpm few second interval. The obtained nano bundle fiber was treated with ethanol to remove excessive water by filtered it using a vacuum pump, to obtain the sheet of nano bundle fiber. Lastly, the fiber underwent ball milling process for 92 h.

2.3 Optimization by Design Expert Software

The experiment design was optimized using Design Expert software, whereas it helps to search for the right factor levels combination, which simulate to satisfy the criteria placed on the responses and factors. There are three factors that are considered for the experiment, which are the amount of polypropylene (wt%), fiber (wt%) and MMT (wt%). The design for response surface was Young's modulus and tensile strength.

2.4 Bamboo Nanofiber Reinforced Polypropylene Composites Preparation

The extracted bamboo nanofiber was reinforced with polypropylene nanocomposite. The ratio of the composites consisted of polypropylene, fiber and MMT were generated from Design Expert software. Table 1 show the composition ratio for eight samples composite.

For each composite, the bamboo nanofiber, polypropylene and MMT were mixed and shake well in transparent zip lock bag based on the ratio generated in Table 1 as shown in Fig. 1. The composites were then placed in the mold and inserted into hot pressed machine for melting and compression at 10 MPa and 180°C for 10 min. It was cold down using cold pressed machine at 27°C for curing before being removed from the mold. The mold was designed according to ASTM D638-14 [20].

Table 1 Ratio of composites composition for eight samples generated by Design Expert software

No of samples	Polypropylene (wt%)	Fibre (wt%)	MMT (wt%)
1	96	3	1
2	93.73	3.28	3
3	93.53	5.47	1
4	91.24	5.76	3
5	90.93	8.07	1
6	88.85	8.15	3
7	88.49	10.51	1
8	85.80	12	2.2

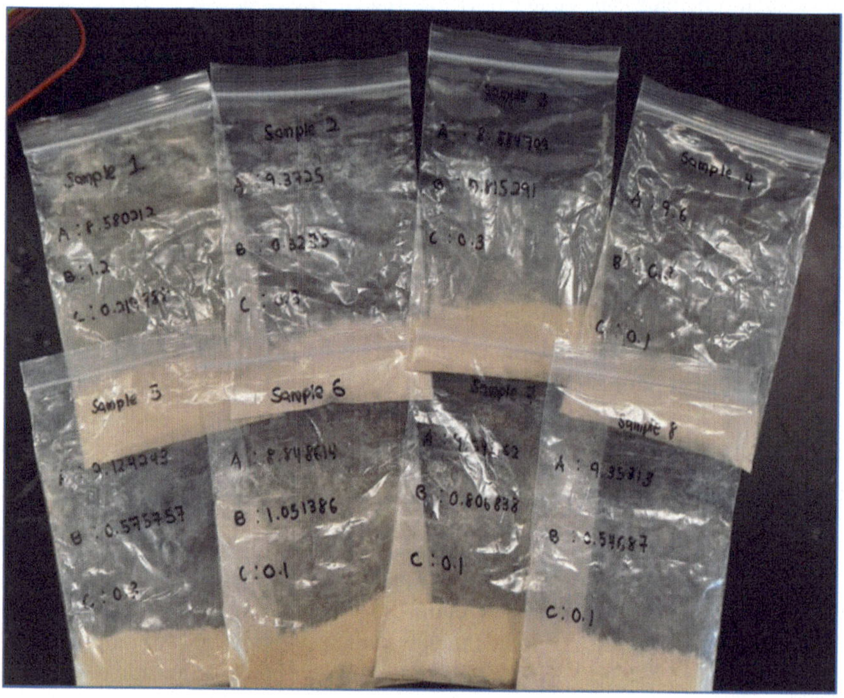

Fig. 1 Eight samples of the composites in zip lock bag

2.5 *Mechanical Test*

The mechanical tests, i.e. tensile tests helped to determine the effectiveness and the behavior of the bamboo fiber reinforce polypropylene when a stretching force applied on it. The maximum strength or load that the composite can withstand was identified. The test can be carried out according to ASTM D638-14 [20] standard using universal testing machine (UTM) (T-machine Technology Machine (Taiwan Co. Ltd.) with the crosshead speed of 5 mm/min.

2.6 *Characterizations*

2.6.1 Fourier Transform Infrared Spectroscopy (FTIR)

FTIR test helped to obtain the infrared spectrum of absorption, emission, photo conductivity or scattering of a solid, liquid or gas. It had an advantage of collecting a high spectral resolution data within a high spectral range. The test of FTIR on the samples were conducted using Shimadzu IR Affinity-1 Spectrophotometer (Shimadzu; Kyoto, Japan) according to ASTM E168-16 [21], and ASTM E1252-98 [22] standards. The range of the wavenumber used was from 4000 to 400 cm^{-1}.

2.6.2 Scanning Electron Microscopy (SEM)

SEM was known as a surface imaging method, whereas the incident electron beams scans across the sample surface and interacts with the sample in order to generate a back-scattered and secondary electron, which are used to create an image of the sample. SEM have an advantage of providing a higher resolution compare to a light microscope as it used electrons that have a smaller wavelength. SEM was done by using a Hitachi TM3030 supplied by JEOL (Tokyo, Japan), with a voltage of 20 kV in a vacuum. The surfaces of the samples are coated and imaged using metal spur coated machine. The tests were conducted according to the ASTM E2015-04 [23] standard.

2.6.3 Energy Dispersive X-Ray Analysis (EDX)

Energy dispersive X-ray analysis (EDX) helped to identify the element exist in the samples and its composition. EDX was carried out the same as SEM method, as it used the same machine, Hitachi TM3030 supplied by JEOL (Tokyo, Japan) with a voltage of 20 kV in a vacuum. However, the tests were conducted according to the ASTM E1508-12 [24] standards

3 Result and Discussion

3.1 Mechanical Properties

Five type of data were obtained from the curve, which consists of tensile strength, yield strength, fracture strength, Young's modulus and elongation at break. The average strength of each composites of samples at peak, yield and break are summarized in Table 2.

3.1.1 Tensile Strength

The results of tensile strength for each samples of the composites were presented in Fig. 2. Based on Fig. 2, Sample 1 consisted polypropylene has the lowest value of tensile. Compared with other samples, the composites loading with fiber and montmorillonite clay (MMT) helped enhanced the tensile strength. The increasing fiber loading effect on the polymer, to increase the tensile strength, which shown in Sample 2, Sample 3, Sample 6 and Sample 7. The improvement of tensile strength increased by 75%, 94%, 10.27% and 80.12% respectively, as compare with pure PP. This was due to the fiber had a higher tensile strength than polymer with rough surface, which produce good interfacial bonding between matrix. Xie et al. [25]

Table 2 The strength for tensile strength, yield strength and fracture strength

Sample	Composition of composites	Tensile strength (MPa)	Yield strength (MPa)	Fracture strength (MPa)
1	(100% PP)	5.5750	2.9391	5.3875
2	(96% PP, 3% Fibre and 1% MMT)	9.7758	5.9602	9.6914
3	(93.73% PP, 3.28% Fibre and 3% MMT)	18.9914	6.4148	18.6586
4	(93.53% PP, 5.47% Fibre and 1% MMT)	14.2312	6.9984	14.0602
5	(91.24% PP, 5.76% Fibre and 3% MMT)	8.0414	5.4773	7.5234
6	(90.93% PP, 8.07% Fibre and 1% MMT)	14.4844	5.5031	14.2148
7	(88.85% PP, 8.15% Fibre and 3% MMT)	16.1414	5.4586	15.9797
8	(88.49% PP, 10.51% Fibre and 1% MMT)	12.4688	7.91953	12.3281
9	(85.80% PP, 12% Fibre and 2.2% MMT)	9.78984	5.39297	9.41719

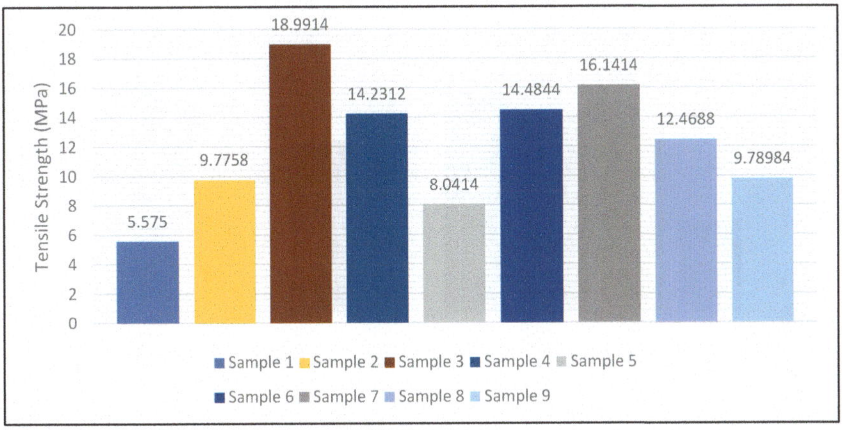

Fig. 2 A comparison of tensile strength for the samples

stated that addition of fiber into polymer helps to increase the stiffness and strength of the composites. However, it was shown that the tensile strength decreases after the maximum or optimum loading of fibers at 5.47 and 8.15%. This was due to lack amount of polymer added, which cannot bind all the bamboo fibers completely. The bamboo fibers irregularity incapable it to support stresses, that were transferred from polymer matrix, which lead to loss of composite strength [26].

The addition of MMT also shows a significant improvement in terms of higher tensile strength, as 3% MMT show a greater percentage of increment compared to 1% MMT. This was due to MMT characteristic, which are high in aspect ratio, small charge density, and big surface area that makes it a suitable filler material for bamboo nanofiber reinforced polypropylene nanocomposites. Rozman et al. [27] reported that the addition of MMT for kenaf fiber-polyester composites showed an improvement in its impact and flexural properties. Hence, it was concluded that optimum fiber loading with MMT in polypropylene is 3.28% and 8.15% respectively both with 3% MMT.

3.1.2 Yield Strength

The yield strength results of each samples of the composites was presented in Fig. 3. From Fig. 3, the highest yield strength was from Sample 8, which consist of 88.49% PP, 10.51% fibers and 3% MMT. The lowest was Samples 1, which only consist of pure polypropylene. Sample 8 withstood up till 7.9195 MPa, which was about 169% increment, as compared to Sample 1, before permanent deformation no longer able to the keep it original state. Hence, addition of fibers and MMT helped to increase the yield strength of the composites.

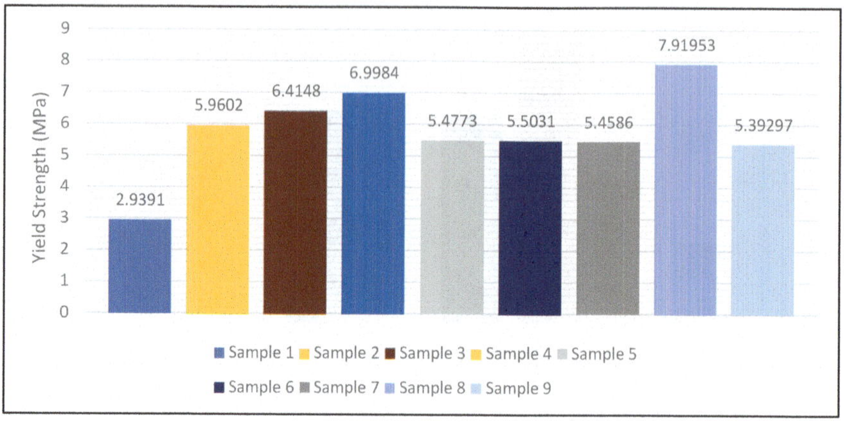

Fig. 3 A comparison of yield strength for the samples

3.1.3 Fracture Strength

The yield strength results for each composite sample was presented in Fig. 4. Based on Fig. 4, compared to pure polypropylenes, all samples except had a high fracture strength. It was proved that the Samples 2 to 9 had the characteristic of ductile materials, as the value of fracture strength were all below than the ultimate tensile strength (UTS). This was supported also by DeGarmo et al. [28]. In a load-controlled case, a ductile material exceeds the maximum tensile strength, and started to deform until the sample rupture. Before ductile material fracture, an extensive plastic deformation or necking happened. This was explained by Callister and Rethwisch [29], whereas extensive plasticity in ductile materials allows the crack to propagate gradually, as it

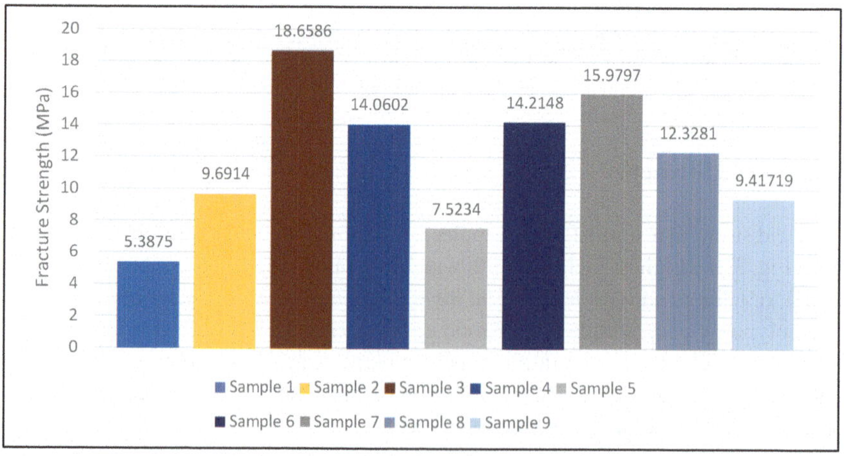

Fig. 4 A comparison of fracture strength for the samples

absorbs huge amount of energy before break. While in brittle material, the fracture strength is the same as UTS. The plastic deformation occurred prior to the fracture. Barsom and Rolfe [30] clarified that brittle fracture required a low energy absorption and happened at a very high velocity, which was up to 7000 feet per second. In addition, Campbell [31] also claimed that brittle material commonly continues to fracture even after the loading was stopped.

3.1.4 Elongation at Breaks

The average percent of elongation at break for the samples was presented in Table 3. The results of elongation at break for each samples of the composites were presented in Fig. 5. Figure 5 show the average percentage of elongation at break was 139.69% for Sample 1 (100% PP). The percentage elongation decrease drastically after Sample 3 (93.73% PP, 3.28% Fibers, 3% MMT) and increase back again in Sample 6 (90.93% PP, 8.07% Fibres, 1% MMT), as that elongation are optimal between the value the range of loading fiber, which was more than 3% till 3.28% and more than 8.07% till 10.51%, respectively. The highest elongation was Sample 8 followed by Sample 3 and Sample 9 with elongation of 182.42%, 181.57% and 179.85%, respectively. This indicated that the three samples had the best ductility, as it has the highest value of percentage elongation. According to Budynas and Nisbett [32], the ductility of material must have an elongation to failure of at least 5% for it to be a significant result. Hence, it can be concluded, the samples were all ductile materials.

Table 3 The average percent of elongation at break for the samples

Sample	Compositions	Elongation at break (%)
1	(100% PP)	139.69
2	(96% PP, 3% Fibre and 1% MMT)	137.72
3	(93.73% PP, 3.28% Fibre and 3% MMT)	181.57
4	(93.53% PP, 5.47% Fibre and 1% MMT)	149.86
5	(91.24% PP, 5.76% Fibre and 3% MMT)	142.81
6	(90.93% PP, 8.07% Fibre and 1% MMT)	127.61
7	(88.85% PP, 8.15% Fibre and 3% MMT)	174.42
8	(88.49% PP, 10.51% Fibre and 1% MMT)	182.42
9	(85.80% PP, 12% Fibre and 2.2% MMT)	179.85

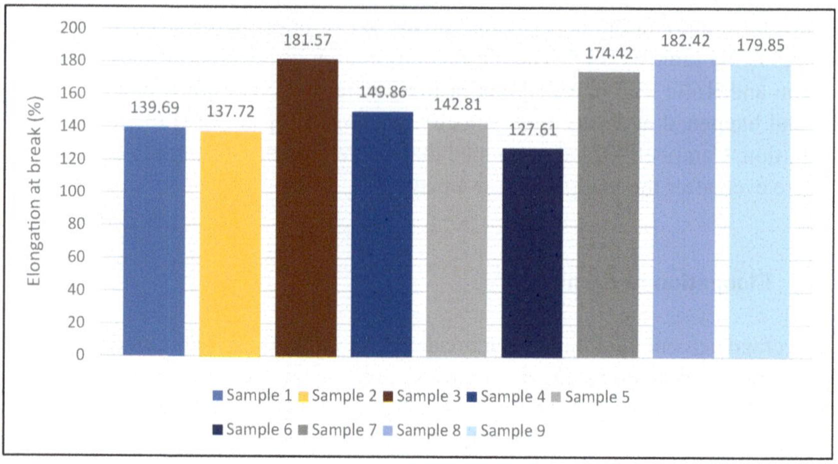

Fig. 5 A comparison of elongation at breaks for the samples

3.1.5 Young's Modulus

The Young's Modulus for the samples was presented in Table 4. The results of Young's modulus for each samples of the composites were presented in Fig. 6. Based on Fig. 6, it was shown that Sample 1 (100% PP) had the lowest value of tensile modulus of 0.4040GPa, whereas the highest value of tensile modulus was 1.0833GPa by Sample 7 (88.85% PP, 8.15% Fibre and 3% MMT). The highest tensile modulus value of the sample was still very small and more towards the region of flexibility. This was proved by Parsons and Goodall [33], whereas it shows ranking approximation for different types of materials in term of strength, toughness and stiffness (Young's Modulus). Based on Fig. 6, the addition of fibers and MMT into the pure polypropylene led to increase of tensile modulus for the composite. The

Table 4 Young's modulus value for the samples

Sample	Compositions	σ (MPa)	ε	E (GPa)
1	(100% PP)	5.5750	0.0138	0.4040
2	(96% PP, 3% Fibre and 1% MMT)	9.7758	0.0142	0.6884
3	(93.73% PP, 3.28% Fibre and 3% MMT)	18.9914	0.0179	1.0610
4	(93.53% PP, 5.47% Fibre and 1% MMT)	14.2312	0.0179	0.7950
5	(91.24% PP, 5.76% Fibre and 3% MMT)	8.0414	0.0125	0.6433
6	(90.93% PP, 8.07% Fibre and 1% MMT)	14.4844	0.0180	0.8047
7	(88.85% PP, 8.15% Fibre and 3% MMT)	16.1414	0.0149	1.0833
8	(88.49% PP, 10.51% Fibre and 1% MMT)	12.4688	0.0173	0.7207
9	(85.80% PP, 12% Fibre and 2.2% MMT)	9.7898	0.0136	0.7198

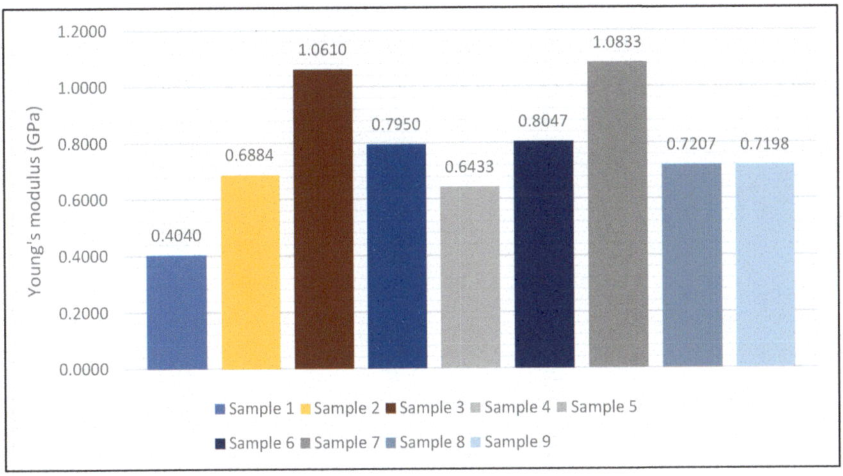

Fig. 6 A comparison of Young's modulus at breaks for the samples

higher the tensile modulus, the higher the elasticity of the materials. Nevertheless, the ductile material was not elastic type, rather a type of plastic, which were more to a soft material. Thus, it was concluded that the lower tensile modulus, the higher the ductility of the materials. All the composite's samples had a low tensile modulus, which was ductile materials.

3.2 Fourier Transform Infrared Spectroscopy (FTIR) Analysis

Figures 7, 8, and 9 show the FTIR graph for raw bamboo, pure polypropylene and Sample 3 (93.73% PP, 3.28% Fibre and 3% MMT), respectively. Sample 3 was chosen due to having the highest value of tensile strength. The 3743.83 cm^{-1} peak Fig. 7 indicated the alkaline C-H stretch. The 2162.20 cm^{-1} peak shows the existence of triple bond between $C \equiv H$ or $C \equiv C$. This happened due to the chemical treatment conducted on the raw bamboo. As for Fig. 8, the 1751.36 cm^{-1} peak indicated double bond of $C = O$ in the polypropylene. For Fig. 10, the highest 1751.36 cm^{-1} peak indicated double bond $C = O$ in the polypropylene the same as in pure polypropylene. This was supported by Merlic et al. [34].

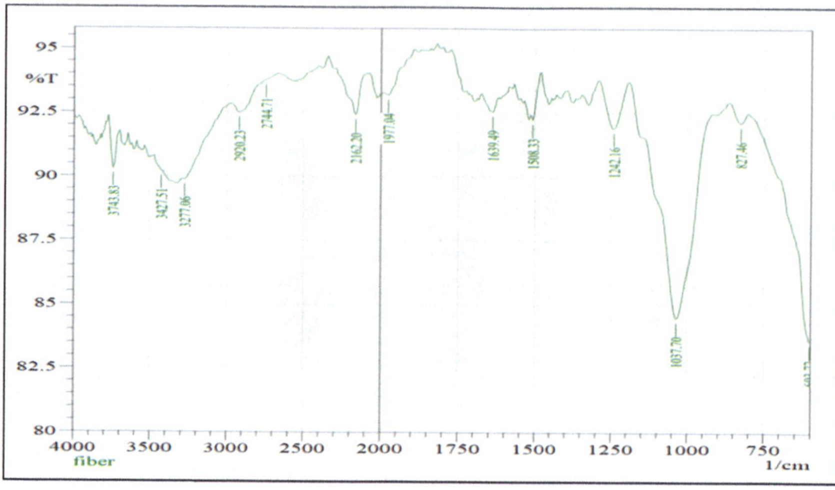

Fig. 7 FTIR graph for raw bamboo

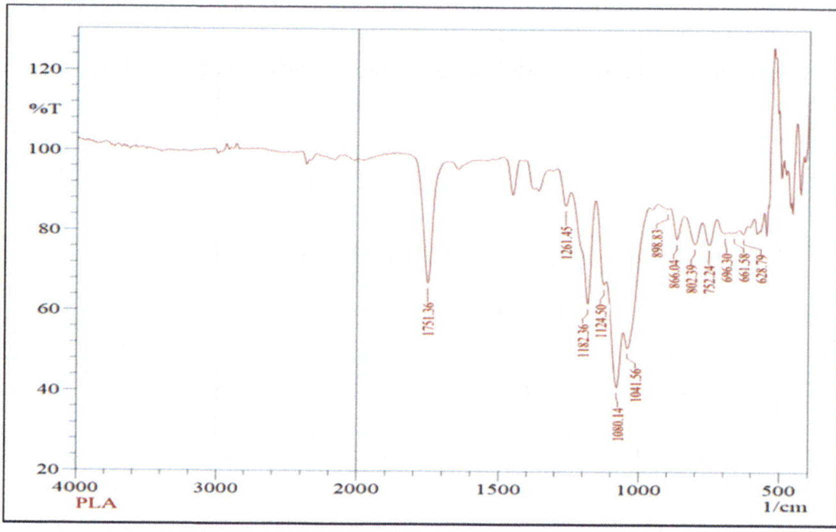

Fig. 8 FTIR graph for sample 1 (PP 100%)

3.3 *Morphological Analysis*

Figure 10a, b show the SEM image of sample 1 (100% PP) in magnification of 100 and 300, respectively. Figure 11a, b show the SEM image of raw bamboo in magnification of 100 and 300, respectively. Figure 12a, b show the SEM image of treated raw bamboo in 5% of sodium hydroxide (NaOH) in magnification of 100

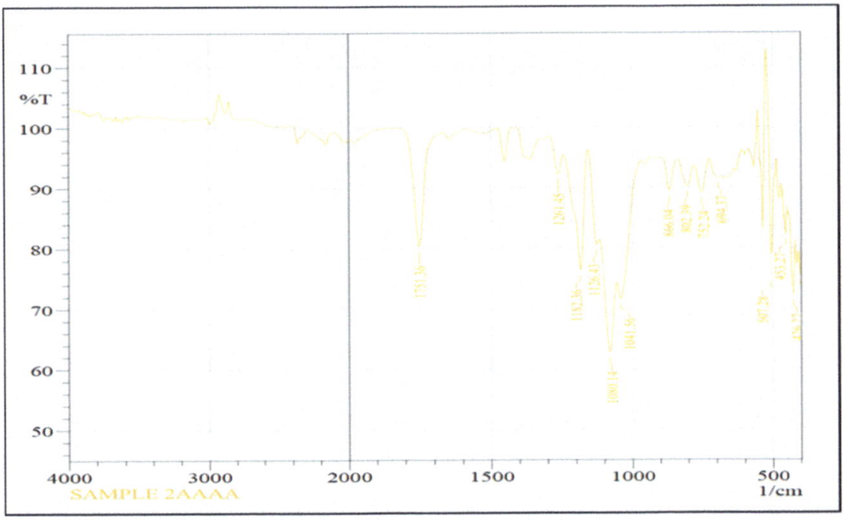

Fig. 9 FTIR graph for Sample 3 (93.73% PP, 3.28% Fibre and 3% MMT)

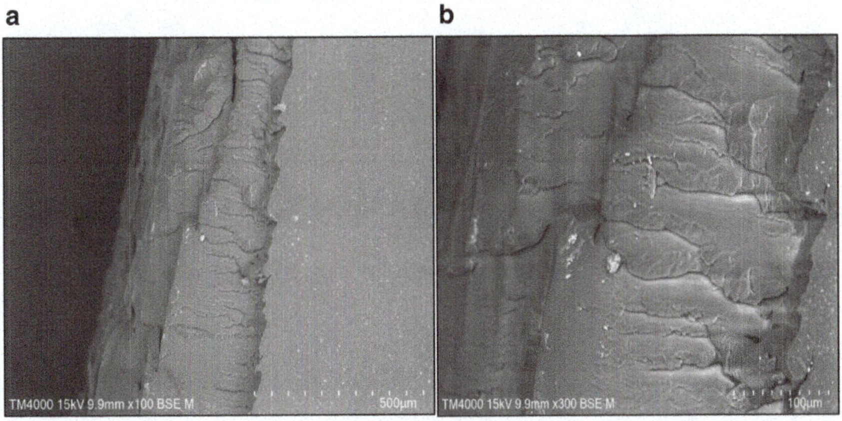

Fig. 10 SEM image of Sample 1 (100% PP) **a** magnification of 100, and **b** magnification of 300

and 300, respectively. Figure 13a, b show the SEM image of treated raw bamboo after milling process of 96 h (4 days) in magnification of 100 and 300 respectively. Figure 14a, b show the SEM image of sample 3 (93.73% PP, 3.28% Fibre and 3% MMT) which has the highest value of tensile strength in magnification of 100 and 300, respectively.

Figure 10a, b shows crack at side of the PP. Both Fig. 11a, b illustrated the morphology of untreated raw bamboo fiber surface. It was observed that the presents of cavities are covered by layer of residues that are due to compression of fibrils. The residue helped protect the fibers and prevent resin penetration through it. According

Fig. 11 SEM image of raw bamboo **a** magnification of 100, and **b** magnification of 300

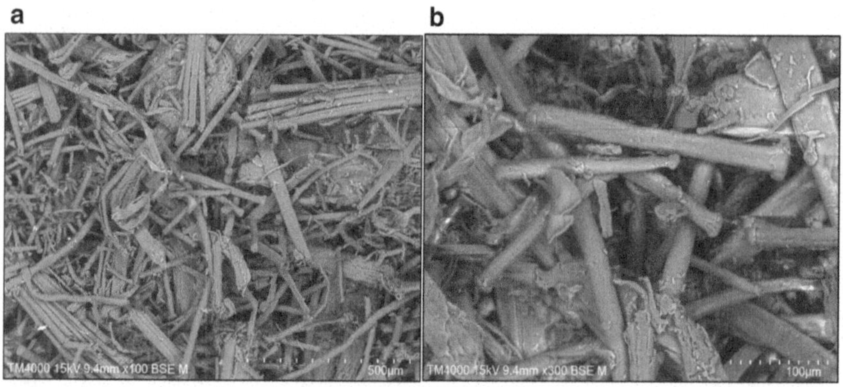

Fig. 12 SEM image of treated raw bamboo **a** magnification of 100, and **b** magnification of 300

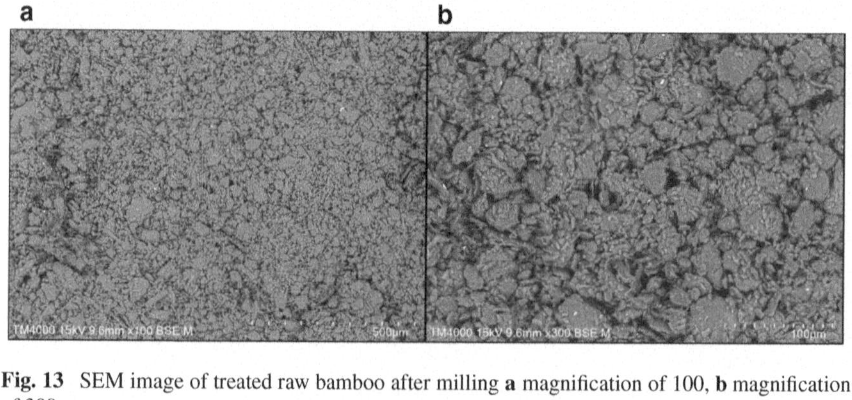

Fig. 13 SEM image of treated raw bamboo after milling **a** magnification of 100, **b** magnification of 300

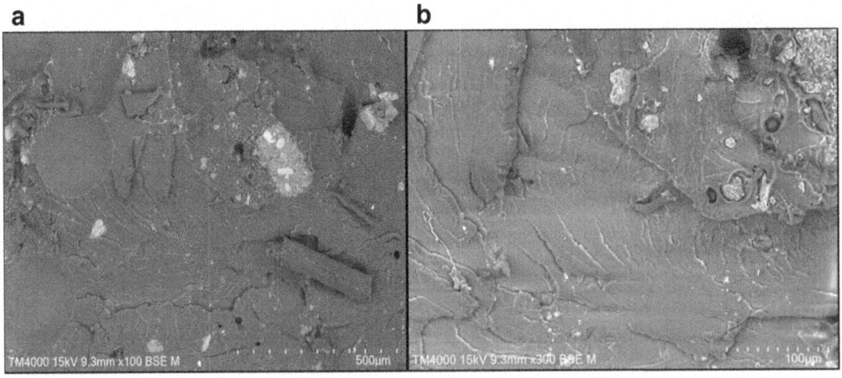

Fig. 14 SEM image of Sample 3 **a** magnification of 100, and **b** magnification of 300

to Alvarez and Vazquez [35], the residues that covered the fibers was also known as aliphatic wax. The aliphatic wax prevents the adhesion between matrix. As for Fig. 12a, b, it was observed that removal of non-cellulose component that consist of lignin, pectin, hemicellulose and waxes, which helps to provide a better bonding in polymer matrix. This was supported by Rout et al. [36], whereas delignification helps to expose the fiber surface more for interaction. Figure 13a, b show much smaller fibers after milling, which helped to the huge increment of surface area for matrix adhesion. Lastly, Fig. 14a, b shows the low interaction of bamboo fibers reinforced polypropylene nano composites based on the limited amount of microfibrillar exposed.

3.4 *Energy Dispersive X-Ray Analysis (EDX) Analysis*

Table 5 show the element and composition exist in sample 1 (100% PP) obtained from EDX analysis. Table 6 show the element and composition exist in treated raw bamboo with treatment of NaOH 5%. Table 7 show the element and composition exist in sample 3 (93.73% PP, 3.28% Fibre and 3% MMT), which was the highest tensile strength value among the samples. Based on Tables 5, 6, and 7, it was carbon had the highest mass composition in Sample 1, followed by Sample 3, which is the

Table 5 Element and composition exist in Sample 1 (100% PP)

Element	Mass composition (%)
Carbon	54.73
Oxygen	44.05
Aluminium	0.74
Fluorine	0.48
Total	100

Table 6 Element and composition exist in treated raw bamboo

Element	Mass composition (%)
Carbon	52.85
Oxygen	47.15
Total	100

Table 7 Element and composition exist in Sample 3

Element	Mass composition (%)
Carbon	53.15
Oxygen	42.77
Nickel	2.67
Silicon	0.86
Iron	0.56
Total	100

composite of polypropylene, fiber and MMT. Sample 3 show the lowest amount of oxygen composition, which was due to the process hot pressing and addition of other fillers.

3.5 Design Expert Optimization

An optimization design was done by using software Design Expert. It helps to search for the best combination factor levels, whereas it simultaneously able to satisfy the criteria placed on the responses and factors. There were factors that considered for the experiment, the amount of polypropylene (wt%) (A), fibre (wt%) (B) and MMT (wt%) (C). The design response surface were tensile modulus and tensile strength. Table 8 show the mixture components used in the design, whereas Table 9 show the responses for the design. There were eight design runs conducted for given composition, which were generated based on the minimum and maximum value set in mixture components. Table 10 show the design layout for the whole experiment, which consist of the composition for the composites, and the value of tensile modulus

Table 8 Mixture components used in the design

Component	Name	Units	Type	Minimum	Maximum
A	Polypropylene	wt%	Mixture	85.8021	96
B	Fibre	wt%	Mixture	3	12
D	MMT	wt%	Mixture	1	3
				Total =	100.00

Table 9 Surface responses used in the design

Response	Name	Units	Analysis	Minimum	Maximum
R1	Tensile Modulus	GPa	Polynomial	0.6433	1.0833
R2	Tensile Strength	MPa	Polynomial	8.0414	18.9914

Table 10 Design layout for the experiment

No of Run	Component 1 A: Polypropylene wt%	Component 2 B: Fibre wt%	Component 3 C: MMT wt%	Response 1 Tensile modulus GPa	Response 2 Tensile strength MPa
1	96	3	1	0.6884	9.7758
2	93.73	3.28	3	1.0610	18.9914
3	93.53	5.47	1	0.7950	14.2312
4	91.24	5.76	3	0.6433	8.0414
5	90.93	8.07	1	0.8047	14.4844
6	88.85	8.15	3	1.0833	16.1414
7	88.49	10.51	1	0.7207	12.4688
8	85.80	12	2.2	0.7198	9.7898

and tensile strength, that have been obtained and calculated after conducting the actual experiment.

Figure 15 show the analysis of variance (ANOVA) for tensile modulus obtained from the design expert where the F-values and P-values were calculated to determine the significance of the model.

Based Fig. 15, the F-value model of 326939.22 implied the model was significant. There was only a 0.01% chance that an F-value this large occurred due to noise. P-values less than 0.0500 indicated model terms are significant. In this case A, B, C, AB, AC, BC, ABC were significant model terms. Values greater than 0.1000 indicated the model terms are less significant. Figure 16 show the coded equation generated by the software for tensile modulus. The equation in terms of coded factors were used to make predictions about the response for given levels of each factor. By default, the high mixture components levels were coded as +1 and the low levels are coded as 0. The coded equation was useful for identifying the relative impact of the factors, by comparing the factor coefficients. Figure 17 show the analysis of variance (ANOVA) for tensile strength obtained from the Design Expert software, whereas the F-values and P-values were calculated to determine the significance of the model. Based on Fig. 17, the model F-value of 435.50 implied the model was significant. There was only a 0.01% chance that an F-value this large could occurred due to noise. P-values less than 0.0500 indicated model terms are significant. In this case A, B, C, AB, AC, BC, ABC were significant model terms. Values greater than 0.1000 indicated the model terms were not significant. Figure 18 show the coded equation generated by the software for tensile strength. Figure 19 show the numerical optimization obtained

ANOVA for Special Cubic model

Response 1: Tensile Modulus

Source	Sum of Squares	df	Mean Square	F-value	p-value	
Model	0.2219	6	0.0370	3.269E+05	< 0.0001	significant
[1]Linear Mixture	0.0489	2	0.0245	2.163E+05	< 0.0001	
AB	0.0135	1	0.0135	1.193E+05	< 0.0001	
AC	0.1123	1	0.1123	9.931E+05	< 0.0001	
BC	0.1064	1	0.1064	9.407E+05	< 0.0001	
ABC	0.1362	1	0.1362	1.204E+06	< 0.0001	
Residual	6.786E-07	6	1.131E-07			
Lack of Fit	6.786E-07	1	6.786E-07			
Pure Error	0.0000	5	0.0000			
Cor Total	0.2219	12				

Fig. 15 ANOVA for response 1: Tensile modulus

Final Equation in Terms of L_Pseudo Components

Tensile Modulus	=
+0.6885	* A
+0.4418	* B
+136.77	* C
+0.9260	* AB
-163.14	* AC
-145.91	* BC
-53.11	* ABC

Fig. 16 Coded equation for tensile modulus

from Design Expert after running 82 runs, whereby the ratio mixture for the three components were obtained in order to get the optimal value of tensile strength and tensile modulus. Based on the Fig. 19 the ratio of the components and the response surface were summarized in Table 11.

ANOVA for Special Cubic model

Response 2: Tensile Strength

Source	Sum of Squares	df	Mean Square	F-value	p-value	
Model	126.67	6	21.11	435.50	< 0.0001	significant
(1)Linear Mixture	7.19	2	3.60	74.19	< 0.0001	
AB	14.57	1	14.57	300.58	< 0.0001	
AC	46.78	1	46.78	965.09	< 0.0001	
BC	44.57	1	44.57	919.38	< 0.0001	
ABC	73.11	1	73.11	1508.05	< 0.0001	
Residual	0.2909	6	0.0485			
Lack of Fit	0.2909	1	0.2909			
Pure Error	0.0000	5	0.0000			
Cor Total	126.96	12				

Fig. 17 ANOVA for response 2: Tensile strength

Final Equation in Terms of L_Pseudo Components

Tensile Modulus	=
+0.6885	* A
+0.4418	* B
+136.77	* C
+0.9260	* AB
-163.14	* AC
-145.91	* BC
-53.11	* ABC

Fig. 18 Coded equation for tensile strength

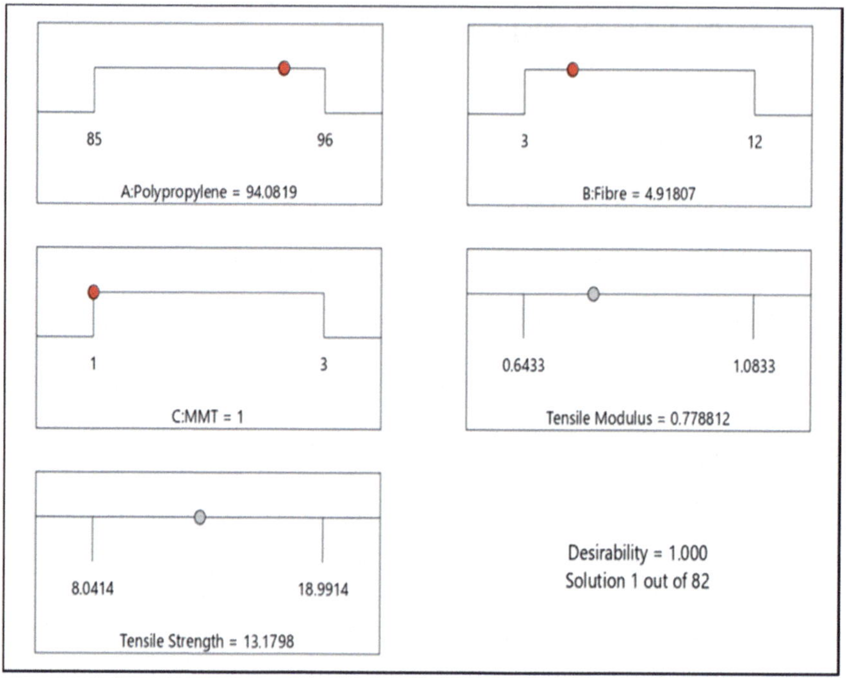

Fig. 19 Optimum ratio for composites

Table 11 Summarization of optimum ratio

Component 1 A: Polypropylene (wt%)	Component 2 B: Fiber (wt%)	Component 3 C: MMT (wt%)	Response 1 Tensile modulus (GPa)	Response 2 Tensile strength (MPa)	Desirability
94.0819	4.91807	1	0.778812	13.1798	1.000
Total = 100 wt%					

4 Conclusions

In conclusion, based on the analyzation of the results, it was proved the addition of polypropylene and MMT as a filler into bamboo fiber help to change the characteristic of the bamboo nanofiber reinforced polypropylene composite from brittle to ductile. Compared to pure polymer, there was also a significant improvement in tensile strength, yield strength, fracture strength and young's modulus of the composites after being reinforced with bamboo nanofiber. The results showed an improvement in tensile strength in the range of 10.27% to 80.12%. The materials ductility was calculated alongside the elongation at break, whereas all the composite samples had a high elongation more than 100, which was in range of elongation between 127.61 and 182.42%. The improvement of ductility was by 3 to 42%. The best Sample 3 consist of 93.73% of polypropylene, 3.28% of fiber and 3% of MMT as the filler, which gave the highest tensile strength of 18.9914 MPa and second highest ductility of 181.57%. An optimization using Design Expert software was carried out to identify the significance of the study, whereas the P-value was less than 0.05. Hence, the assumption of inserting polypropylene and MMT into bamboo fiber helped to increase the tensile strength and tensile modulus.

Acknowledgements The authors would like to acknowledge Universiti Malaysia Sarawak (UNIMAS) for the support.

References

1. Prakash, C.: Bamboo fibre. In: Kozlowski, R. M., Mackiewicz-Talarczyk, M. (eds.), Handbook of natural fibres, pp. 219–229. Woodhead Publishing, Cambridge (2020). https://doi.org/10.1016/B978-0-12-818398-4.00009-8
2. Ibrahim, I.D., Jamiru, T., Sadiku, R.E., Kupolati, W.K., Agwuncha, S.C., Ekundayo, G.: The use of polypropylene in bamboo fibre composites and their mechanical properties—a review. J. Reinf. Plast. Compos. **34**(15), 1347–1356 (2015). https://doi.org/10.1177/0731684415591302
3. Liu, D., Song, J., Anderson, D.P., Chang, P.R., Hua, Y.: Bamboo fiber and its reinforced composites: structure and properties. Cellulose **19**(1), 1449–1480 (2012). https://doi.org/10.1007/s10570-012-9741-1
4. Pickering, K.L., Aruan Efendy, M.G., Le, T.M.: A review of recent developments in natural fibre composites and their mechanical performance. Compos. Part A: Appl. Sci. Manuf. **83**(1), 98–112 (2016). https://doi.org/10.1016/j.compositesa.2015.08.038
5. Kalia, S., Kaith, B.S., Kaur, I.: Pretreatments of natural fibers and their application as reinforcing material in polymer composites—A review. Polym. Eng. & Sci. **49**(7), 1253–1272 (2009). https://doi.org/10.1002/pen.21328
6. Rajak, D.K., Pagsr, D.D., Menezes, P.L., Linul, E.: Fiber-reinforced polymer composites: manufacturing, properties, and applications. Polymers **11**(1), 1667 (2019). https://doi.org/10.3390/polym11101667
7. Okubo, K., Fujii, T., Yamamoto, Y.: Development of bamboo-based polymer composites and their mechanical properties. Compos. Part A: Appl. Sci. Manuf. **35**(3), 377–383 (2004). https://doi.org/10.1016/j.compositesa.2003.09.017
8. Imadi, S.R, Mahmood, I., Kazi, A.G.: Bamboo fiber processing, properties, and applications. In: Hakeem, K.R., Jawaid, M., Rashid, U. (eds.), Biomass and bioenergy, pp. 27–46, Springer, Cham (2014). https://doi.org/10.1007/978-3-319-07641-6_2
9. Cao, Y., Wu, Y.-Q.: Evaluation of statistical strength of bamboo fiber and mechanical properties of fiber reinforced green composites. J. Cent. South Univ. Technol. **15**(1), 564–567 (2010). https://doi.org/10.1007/s11771-008-0422-z
10. Liu, L., Wang, Q., Cheng, L., Qian, J., Yu, J.: Modification of natural bamboo fibers for textile applications. Fibers Polym. **12**(1), 95–103 (2011). https://doi.org/10.1007/s12221-011-0095-3
11. Hakeem, K.R., Ibrahim, S., Ibrahim, F.H., Tombuloglu, H.: Bamboo biomass: various studies and potential applications for value-added products. In: Hakeem, K., Jawaid, M.Y., Alothman, O. (eds), Agricultural biomass based potential materials, pp. 231–243. Springer, Cham (2015). https://doi.org/10.1007
12. Adamu, M., Rahman, M.R., Hamdan, S., Bakri, M.K.B., Md Yusof, F.A.B.: Impact of polyvinyl alcohol/acrylonitrile on bamboo nanocomposite and optimization of mechanical performance

by response surface methodology. Constr. Build. Mater. **258**(1), 119693 (2020). https://doi.org/10.1016/j.conbuildmat.2020.119693
13. Andrady, A.L., Neal, M.A.: Applications and societal benefits of plastics. Philos. Trans. R. Soc. B: Biol. Sci. **364**(1526), 1977–1984 (2009). https://doi.org/10.1098/rstb.2008.0304
14. Zaaba, N.F., Ismail, H. (2019). Thermoplastic/natural filler composites: a short review. J. Phys. Sci. **30**(1), 81–99. https://doi.org/10.21315/jps2019.30.s1.5
15. Maddah, H.A.: Polypropylene as a promising plastic: a review. Am. J. Polym. Sci. **6**(1), 1–11 (2016). https://doi.org/10.5923/j.ajps.20160601.01
16. Todkar, S.S., Patil, S.A.: Review on mechanical properties evaluation of pineapple leaf fibre (PALF) reinforced polymer composites. Compos. Part B: Eng. **174**(1), 106927 (2019). https://doi.org/10.1016/j.compositesb.2019.106927
17. Javadian, A., Smith, I.F.C., Saeidi, N., Hebel, D.E.: Mechanical properties of bamboo through measurement of culm physical properties for composite fabrication of structural concrete reinforcement. Front. Mater.: Mech. Mater. **6**(1), 1–15 (2019). https://doi.org/10.3389/fmats.2019.00015
18. Onikeku, O., Shitote, S.M., Mwero, J., Adedeji, A.A.: Evaluation of characteristics of concrete mixed with bamboo leaf ash. Open Constrution & Build. Technol. J. **13**(1), 67–80 (2019). https://doi.org/10.2174/1874836801913010067
19. Karatas, M.A., Gokkaya, H.: A review on machinability of carbon fiber reinforced polymer (CFRP) and glass fiber reinforced polymer (GFRP) composite materials. Def. Technol. **14**(4), 318–326 (2018). https://doi.org/10.1016/j.dt.2018.02.001
20. ASTM D638-14: Standard test method for tensile properties of plastics. ASTM International, West Conshohocken, PA (2014). https://doi.org/10.1520/D0638-14
21. ASTM E168-16: Standard practices for general techniques of infrared quantitative analysis. ASTM International, West Conshohocken, PA (2016). https://doi.org/10.1520/E0168-16
22. ASTM E1252-98: Standard practice for general techniques for obtaining infrared spectra for qualitative analysis. ASTM International, West Conshohocken, PA (2013). https://doi.org/10.1520/E1252-98R13E01
23. ASTM E2015-04: Standard guide for preparation of plastics and polymeric specimens for microstructural examination. ASTM International, West Conshohocken, PA (2014). https://doi.org/10.1520/E2015-04R14
24. ASTM E1508-12: Standard guide for quantitative analysis by energy-dispersive spectroscopy. ASTM International, West Conshohocken, PA (2019). https://doi.org/10.1520/E1508-12AR19
25. Xie, Y., Hill, C.A.S., Xiao, Z., Militz, H., Mai, C.: Silane coupling agents used for natural fiber or polymer composites: a review. Compos. Part A: Appl. Sci. Manuf. **41**(7), 806–819 (2010). https://doi.org/10.1016/j.compositesa.2010.03.005
26. Facca, A.G., Kortschot, M.T., Yan, N.: Predicting the tensile strength of natural fibre reinforced thermoplastics. Compos. Sci. Technol. **67**(11–12), 2454–2466 (2007). https://doi.org/10.1016/j.compscitech.2006.12.018
27. Rozman, H.D., Rozyanty, A.R., Musa, L., Tay, G.S.: Ultra-violet radiation cured biofiber composites from kenaf: the effect of montmorillonite on the flexural and impact properties. J. Wood Chem. Technol. **30**(2), 52–163 (2010). https://doi.org/10.1080/02773810903259565
28. Black, J.T., Kohser, R.A.: Degarmo's materials and process in manufacturing. Wiley (2003)
29. Callister, W.D., Rethwisch, D.G.: Materials science and engineering: an introduction. Wiley (2018)
30. Barsom, J.H., Rolfe, S.T.: Fracture and fatigue control in structures: applications of fracture mechanics. ASTM International, West Conshohocken (1999)
31. Campbell, F.C.: Fatigue and fracture: understanding the basics. ASM International (2012)
32. Budynas, R.G., Nisbett, J.K.: Shigley's mechanical engineering design. McGraw-Hill Education (2015)
33. Parsons, L., Goodall, R.: Testing the fracture behaviour of chocolate. Phys. Educ. **45**(50), 1–5 (2010). https://doi.org/10.1088/0031-9120/46/1/006
34. Merlic, C.A., Fam, B.C., Miller, M.M.: Webspectra: online NMR and IR spectra for students. J. Chem. Educ. **78**(1), 118 (2001). https://doi.org/10.1021/ed078p118

35. Alvarez, V.A., Vazquez, A.: Thermal degradation of cellulose derivatives/starch blends and sisal fibre composites. Polym. Degrad. Stab. **84**(1), 13–21 (2004). https://doi.org/10.1016/j.polymdegradstab.2003.09.003
36. Rout, J., Tripathy, S.S., Nayak, S.K., Misra, M., Mohanty, A.K.: Scanning electron microscopy study of chemically modified coir fibers. J. Appl. Polym. **79**(7), 1169–1177 (2000). https://doi.org/10.1002/1097-4628(20010214)79:7%3c1169:AID-APP30%3e3.0.CO;2-Q

Bamboo and Wood Fibers/MMT Hybrid Nanocomposites

Md Rezaur Rahman, Muhammad Khusairy Bin Bakri, and Sinin Hamdan

Abstract In this research, bamboo and wood fibers were used a reinforcement to fabricate the nanocomposite, while montmorillonite (MMT) is used as nanofiller. Bamboo and wood fiber contain constituents of cellulose, which had a high strength property. However, the hydroxyl group attached on to the cellulose chain made the bamboo and wood fiber hydrophilic, which weaken the interfacial bonding. Henceforth, bamboo and wood fiber surface treatment were needed to improve its morphological and mechanical properties. Surface treatment was done before the fabrication of the nanocomposite, which was during the fiber's extraction process. For the fabrication of hybrid nanocomposite process, Design Expert software was used to determine the optimum MMT, bamboo, and wood fiber ratio, which was known as the response surface methodology (RSM). The highest modulus of rapture (MOR) value was achieved at 15.8008 MPa for the nanocomposite with the weight percentage of 7wt% wood fiber, 7wt% bamboo fiber, 10wt% MMT and 76wt% PLA. As for the modulus of elasticity (MOE), the highest value obtained was at 0.08 GPa, which has the weight percentage of 6.03wt% for wood fiber, 4.5wt% for bamboo fiber, 1wt% for MMT and 88.47wt% for PLA. The morphological properties of the nanocomposites were studied by using scanning electron microscopy (SEM), energy dispersive x-ray (EDX), and Fourier transform infrared spectroscopy (FTIR) equipment, which illustrated the interfacial reaction that occur in the nanocomposite.

Keywords Bamboo · MMT · Woods · Nanocomposites · Optimization

1 Introduction

In the face of the introduction green technology, which begun a few years ago, it is not until now that the environmental issue had gained the attention of every corner of the world [1]. The technology advancement on environmentally friendly approaches had grown bigger, even though it was still in the starting phase [2]. The best approach

M. R. Rahman (✉) · M. K. B. Bakri · S. Hamdan
Faculty of Engineering, Universiti Malaysia Sarawak, Jalan Datuk Mohammad Musa, 94300, Kota Samarahan, Sarawak, Malaysia
e-mail: rmrezaur@unimas.my

M. R. Rahman (ed.), *Bamboo Polymer Nanocomposites*, Engineering Materials,
https://doi.org/10.1007/978-3-030-68090-9_6

towards sustainability was to grow own resources, which give less impact to the environment [3]. In terms of bio-composites development that are made from natural fibers, the approach certainly going towards the eco-environment friendly [4].

Categorized as green materials, bamboo fiber as bio-resources had gained popularity among researchers, due to its characteristics, which replace popular synthetics fiber, i.e. polyester [5]. Bamboo was found growing abundantly in Asian region, which usually used as food source, decoration, craft furniture, a building material, and versatile raw product [5]. In composite material development, bamboo was considered highly significance due to its cheap, availability, rapid growth, high strength, and environmentally friendly in nature [6–8]. Therefore, for industrial sector, the high-performance characteristic of bamboo benefited both the material developer and the environment [9–11]. The bamboo cultivation reduced greenhouse gases emission effects that contributes to the global warming [12]. In fact, these problems had become a serious threat situation that made the world concerns about the environment effects today [13]. In Malaysia, several types of bamboo available, i.e. buluh galah (*bambusa heterostachya*), buluh beting (*dendrocalamus asper*), buluh duri (*bambusa blumeana*), buluh tumpat tikus (*gigantochloa ligulate*), buluh betong (*gigantochloa levis*), buluh gading (*bambusa vulgaris*), buluh lemang (*schizostachyum brachycladum*), buluh semantan (*gigantochloa scortechinii*), buluh gading (*bambusa vulgaris varstriata*), buluh dinding (*shizostachyum zollingeri*), buluh beti (*gigantochloa wrayi*), and buluh semenyeh (*schizostachyum grande*) [14]. In construction and building materials, bamboo had both advantages and disadvantages, especially on the use of its bamboo fiber, as the major chemical compositions were cellulose, hemicellulose and lignin [5, 6]. For the advantages, bamboo lignin and cellulose structures provides a high thermal, and stiffness properties, respectively [15–20]. While, the disadvantages of bamboo fiber, it was highly hydrophilic, which had high water uptake/intake [15–20].

Wood fiber also considered as another bio-resource that had similar material properties and utilization [21]. However, as compared to bamboo, wood had a broad of applications, i.e. timber for furniture, building material, veneer, fiberboard, plywood, and other wood-based product [22]. As renewable resource, wood-based composites producing material provided environment friendly benefits many companies [23]. Cellulose, hemicellulose and lignin in wood fibers posed the same advantages and disadvantages of bamboo, i.e. high tensile strength even though it was hydrophilic in nature, which is also its disadvantage. However, the growth rate between wood and bamboo made a huge difference between, especially in harvesting process, whereas wood need more time to grow [22].

Nanoclay was one of an inorganic nanomaterial introduced as filler in the matrix for organic/inorganic nanocomposites fabrication [24]. One of the nanoparticles was montmorillonite (MMT) that acted as nanofiller for polymeric matrices [25]. Based on Garusinghe et al. [26], MMT was known to have an ability to improve both barrier and mechanical properties of the composite in high in-plane strength and stiffness. In addition, MMT has strong adsorption as well as high affinity of several substances, i.e. heavy metal ions and biological materials [27]. In order to ensure the homogeneity of MMT dispersion and produce a strong bond of nanomaterials, a correct dispersion technique is required. For the MMT nanomaterials diffusion, MMT was dispersed, which created a tortuous path that permeated the molecules. For a strong bamboo and wood/MMT hybrid nanocomposite fabrication, to obtain a maximum amount of bamboo fiber, the bamboo and wood fiber must be extracted using the proper technique. Few surface treatments were done onto the materials, by improving its chemical and physical properties, i.e. reducing the bamboo and wood fibers' water uptake, increasing bonding tendency, and increasing the compatibility of the fibers with the nanoclay matrices. Therefore, this chapter discussed the results of the bamboo and wood/MMT hybrid nanocomposites through characterization, i.e. physico-chemical morphological analysis, and mechanical properties.

2 Methodology

2.1 Materials

Raw wood and raw bamboo were obtained from a local bamboo farm and forest in Kampung Daun, Singai, Bau Sarawak. Ethanol, toluene, montmorillonite (MMT), polylactic acid, acetic acid, titanium (IV) oxide, sodium hydroxide and hydrogen peroxide were obtained from Sigma Aldrich (US) Company.

2.2 Response Surface Methodology by Design Expert

Design Expert® 11 software was used for cresting response surface methodology (RSM) model. It was conducted by obtaining the data set values of determine best composition, which provides maximum output of nanocomposites with high strength.

2.3 Bamboo Fiber Extraction

The raw bamboo was cut into small pieces of cubed-shaped (1 cm in width × 1 cm in length × 1 cm in length) and cleaned using distilled water. All physically bound moisture was removed from bamboo by dried it for at 110°C 24 h. A part of the dried raw bamboo was kept in a disclosed plastic container for prior analysis.

By using Soxhlet assembly, 10 g of raw bamboo was extracted using a mixture of 200 mL ethanol and 400 mL toluene to remove the wax, fats, oil, resins, and phenolic extractives. From the Soxhlet extractor, a tweezer was used to pullout the cellulose thimble, whereas the product was poured into a beaker and stirred with glass rod, while ethanol-toluene mixture was added. Bamboo fiber delignification was done by using delignification solution, which was prepared using 82.3 g (35wt%) hydrogen peroxide (H_2O_2) and 106.2 g (99.8wt%) acetic acid (CH_3COOH) in the presence of titanium (IV) oxide catalyst. 30 g of the dried bamboo fiber was immersed into the delignification solution and heated for 2 h at temperature 130°C. As the solution cooled down, the samples were filtered and cleaned using deionized water until it achieved neutral pH7 state. The filtered sample was dried in the oven for 24 h at 70°C.

For the bamboo mercerization process, the sample was immersed into alkaline solution (sodium hydroxide), whereas the mixture in the beaker was placed in the auto shaker and stirred for 2 h at 150 rpm at 80°C. The sample continued to be stirred for 8 h with no heat turned on. Lastly, the sample was filtered and rinsed until it reached pH 7. After this process, the sample was placed in the freezer to freeze-dry it at temperature -85°C for 48 h. Prior to the composite molding, the sample was ground using ball milling machine to reduce it size.

2.4 *Extracting Fiber from Wood*

The raw wood sawdust was collected, cleaned with distilled water, and dried at 110°C for 24 h to removed moisture. A part of the dried sawdust was kept further analysis. 10 g of dry wood was added in 300 mL of deionized water at 70°C. The mixture was kept under constant stirring for 1 h before cooled to room temperature, followed by filtering on coarse membrane filter paper. The wood was dried in an oven for at 50°C 3 h. After that, 10 g of material was refluxed with ethanol for 5 h at 80°C in a cellulose thimble in a Soxhlet assembly. 5 g of samples was mixed with 0.06L of 45 g/L NaOH for alkaline treatment and hydrolyzed at 80°C for 2 h. The reaction was stopped by cooling it with ice bath. The sample was filtered by using thick membrane filter paper. The sample was dried at 50°C in an oven for 5 h. The dried samples were bleached with a mixture of 5 g of wood, 0.1L of 50 g/L NaOH. Then using 0.1L of 160 g/L H_{-2}-O_{-2-}, the samples was hydrolyzed at 55°C for 90 min. The hydrolysate sample was filtered with a medium membrane filter paper and washed with deionized water until it reached neutral pH7. The samples were dried again in the oven at 50°C for 15 h. Prior to the composite fabrication, the sample was ground using ball milling machine to reduce the size.

2.5 *Wood and Bamboo Fiber/MMT Hybrid Nanocomposite Fabrication*

Polylactic acid was is used for the nanocomposite fabrication. The materials were prepared accordingly by referring to the generated composition of nanocomposite sample obtained from the Design Expert® 11 software. Wood fiber, bamboo fiber, MMT and PLA were weighted and mixed. 20 different compositions were prepared

Table 1 Compositions of materials for different nanocomposite sample based on design expert RSM

Run	Component 1 A: Wood (wt%)	Component 2 B: Bamboo (wt%)	Component 3 C: MMT (wt%)	Component 4 D: PLA (wt%)
1	7	4.5	3.187	85.313
2	4.5	4.5	6.85	84.15
3	7	7	10	76
4	5.52876	7	4.9796	82.4916
5	4.5	6.93	3.3503	85.2197
6	7	4.5	7.42459	81.0754
7	4.5	6.05415	1	88.4459
8	4.5	4.5	9.17165	81.8284
9	7	4.5	10	78.5
10	5.69862	5.65713	9.64977	78.9945
11	6.02857	4.5	1	88.4714
12	5.69862	5.65713	9.64977	78.9945
13	4.5	7	10	78.5
14	6.96394	5.55255	4.89852	82.585
15	5.69862	5.65713	9.64977	78.9945
16	7	7	1	85
17	5.52876	7	4.9796	82.4916
18	7	7	6.99281	79.0072
19	6.96394	5.55255	4.89852	82.585
20	6.02857	4.5	1	88.4714

as shown in Table 1. The mixture was placed and hot pressed using hot pressing machine at temperature of 180°C for 15 min. The mold was then left to cool down at room temperature. The mold was done according to ASTM D638-14 [28] standard.

2.6 *Mechanical Test*

The modulus of rupture (MOR) and modulus of elasticity (MOE) were obtained. Universal Testing Machine (UTM) was used with a crosshead speed of 5 mm/min. The result was obtained from the computer and recorded for analysis according to ASTM D638-14 [28] standard.

2.7 *Characterization of Nanocomposite*

2.7.1 Scanning Electron Microscopy (SEM) and Energy Dispersive X-Ray (EDS/EDX)

SEM analysis was performed on the break or ruptured cross sections of the, as well on the bamboo and wood fiber, according to ASTM E2015-04 [29] standard. The samples were mounted on sample plate holder. The samples were placed in the chamber for SEM analysis. Hitachi Analytical Tabletop SEM (benchtop) 'TM-3030' (Hitachi High-Technologies (Germany) Europe GmbH.) was used for the SEM analysis. EDS/EDS was also carried out in combination with the SEM analysis to identify elements found on the surface. It was done according to ASTM E1508-12 [30] standard.

2.7.2 Fourier Transform Infrared (FTIR) Spectroscopy

FTIR spectra was recorded in the wavenumber range of 4000-400 cm^{-1} with a resolution of 1 cm^{-1}. 'IRAffinity-1' spectroscopy (Shimadzu, Japan) machine was used for the FTIR analysis. The tests were done according to ASTM E168-16 [31], and ASTM E1252-98 [32] standards.

3 Results and Discussions

3.1 *Mechanical Properties*

The minimum and maximum composition of each materials were defined, whereas the Design Expert software generates compositions variation of materials based on the defined limits for each material. The variables for the design of the experiment was shown in Table 2. Table 3 shows the generated composites sample with different materials composition, which were tested using a tensile test machine. Based on Table 3, there were 20 samples that were prepared.

Table 2 Variables of composite ingredients for the design of the experiment

Component	Name	Units (%)	Levels	
			Low	High
A	Wood	wt	4.5	7
B	Bamboo	wt	4.5	7
C	MMT	wt	1	10
D	PLA	wt	76	88.4714

Table 3 Composite samples ingredients and the responding modulus of rupture (MOR) and modulus of elasticity (MOE)

Run	Component 1 A: Wood (wt%)	Component 2 B: Bamboo (wt%)	Component 3 C: MMT (wt%)	Component 4 D: PLA (wt%)	Response 1 MOR (MPa)	Response 2 MOE (GPa)
1	7	4.5	3.187	85.313	13.8117	0.0303061
2	4.5	4.5	6.85	84.15	14.5018	0.0107905
3	7	7	10	76	15.8008	0.0257405
4	5.52876	7	4.9796	82.4916	14.5528	0.0100307
5	4.5	6.93	3.3503	85.2197	14.8162	0.0156509
6	7	4.5	7.42459	81.0754	15.012	0.0183703
7	4.5	6.05415	1	88.4459	10.594	0.0110177
8	4.5	4.5	9.17165	81.8284	14.8706	0.0115483
9	7	4.5	10	78.5	15.2141	0.0128882
10	5.69862	5.65713	9.64977	78.9945	15.5803	0.0312569
11	6.02857	4.5	1	88.4714	13.0836	0.0771753
12	5.69862	5.65713	9.64977	78.9945	15.5803	0.0312569
13	4.5	7	10	78.5	15.2011	0.0116233
14	6.96394	5.55255	4.89852	82.585	15.1221	0.0136255
15	5.69862	5.65713	9.64977	78.9945	15.5803	0.0312569
16	7	7	1	85	12.0532	0.0258964
17	5.52876	7	4.9796	82.4916	14.5528	0.0100307
18	7	7	6.99281	79.0072	15.6057	0.0100802
19	6.96394	5.55255	4.89852	82.585	15.1221	0.0136255
20	6.02857	4.5	1	88.4714	11.5711	0.0771753

3.1.1 Modulus of Rupture (MOR)

The modulus of elasticity of the samples were obtained, data analysis was carried out in the Design Expert software using ANOVA. The suggested linear model was chosen, and the resulting ANOVA was shown in Table 4. The fit statistic of the design was also provided, whereas the design has an R^2 value of 0.7369, adjusted R^2 value of 0.6876 and predicted R^2 value of 0.5401, as shown in Table 5.

The linear mixture for the model shows that the MOR is a function of linear mathematical model for the components shown in Eq. (1).

Table 4 ANOVA results for linear model of MOR

Source	Sum of squares	Mean square	F-value	P-value
Model	30.10	10.03	14.94	<0.0001
Linear mixture	30.10	10.03	14.94	<0.0001

Table 5 Fit statistics for MOR model

R^2	0.7369
Adjusted R^2	0.6876
Predicted R^2	0.5401

$$MOR = 0.375523(Wood) + 0.2733378(Bamboo) + 0.4543708(MMT) + 0.0959138(PLA) \tag{1}$$

This model was generated by the Design Expert software, which was used to predict the MOR response at different weight percentage. However, due to the low predicted R^2 value and adjusted R^2 value, the accuracy of the experimental value may less significant. Table 6 below shows the predicted MOR response using the value of the different weight percentage compositions, whereas the model was generated by the software.

Table 6 Comparison between the experimental and predicted results

Run	Component 1 A: Wood (wt%)	Component 2 B: Bamboo (wt%)	Component 3 C: MMT (wt%)	Component 4 D: PLA (wt%)	Experimental MOR (MPa)	Predicted MOR (MPa)
1	7	4.5	3.187	85.313	13.8117	13.4894549
2	4.5	4.5	6.85	84.15	14.5018	14.1034599
3	7	7	10	76	15.8008	16.3751824
4	5.52876	7	4.9796	82.4916	14.5528	14.1642088
5	4.5	6.93	3.3503	85.2197	14.8162	13.2801082
6	7	4.5	7.42459	81.0754	15.012	15.0084477
7	4.5	6.05415	1	88.4459	10.594	12.2822347
8	4.5	4.5	9.17165	81.8284	14.8706	14.9356763
9	7	4.5	10	78.5	15.2141	15.9316224
10	5.69862	5.65713	9.64977	78.9945	15.5803	15.6475067
11	6.02857	4.5	1	88.4714	13.0836	12.4338858
12	5.69862	5.65713	9.64977	78.9945	15.5803	15.6475067
13	4.5	7	10	78.5	15.2011	15.6761594
14	6.96394	5.55255	4.89852	82.585	15.1221	14.2796271
15	5.69862	5.65713	9.64977	78.9945	15.5803	15.6475067
16	7	7	1	85	12.0532	13.1490694
17	5.52876	7	4.9796	82.4916	14.5528	14.1642088
18	7	7	6.99281	79.0072	15.6057	15.2972351
19	6.96394	5.55255	4.89852	82.585	15.1221	14.2796271
20	6.02857	4.5	1	88.4714	11.5711	12.4338858

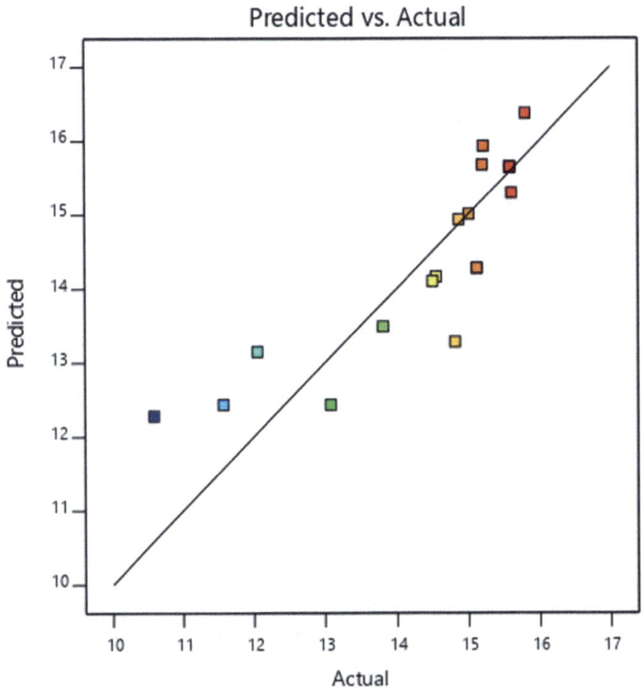

Fig. 1 Predicted vs actual graph for MOR

A predicted vs experimental graph was plotted and shown in Fig. 1. Based on the graph, the predicted and the actual value scatters along the middle line. Few factors led to the result inaccuracy, which come from the composite itself. Since the model was created based on the response of the MOR obtained from the experimental work, a faulty composite sample misled from the accurate value lead to a low R^2, adjusted R^2 and predicted R^2 value, cause less accurate model. However, the difference between the predicted R^2 value of 0.5401 and the adjusted R^2 value of 0.6876, was less than 0.2, making the model somewhat acceptable as denoted by the Design Expert software. Based on Adamu et al. [33], the predicted R^2 values shows the developed model generate response for new observations that has different compositions weight percentage (wt%) than the generated value as shown in Table 3. It also determines whether the model is complicated or otherwise.

3.1.2 Modulus of Elasticity (MOE)

A similar approach was conducted for the MOE with the response shown in Table 3. Based on the result, the following ANOVA results was obtained, which was shown in Table 7. While the fit statistic was shown in Table 8, which has the R^2 value of 0.9140, adjusted R^2 value of 0.8366 and predicted R^2 value of 0.2884.

Table 7 ANOVA results for quadratic model of MOE

Source	Sum of squares	Mean square	F-value	P-value
Model	0.0069	0.0008	11.81	0.0003
Linear mixture	0.0024	0.0008	12.20	0.0011
AB	0.0004	0.0004	6.34	0.0305
AC	0.0008	0.0008	12.90	0.0049
AD	0.0011	0.0011	17.44	0.0019
BC	0.0000	0.0000	0.4663	0.5102
BD	0.0001	0.0001	1.19	0.3006
CD	0.0010	0.0010	15.76	0.0026

Table 8 Fit statistics for MOE model

R^2	0.9140
Adjusted R^2	0.8366
Predicted R^2	0.2884

By referring to the ANOVA, the P-values of some of the model terms was less than 0.0500, which means that these terms were significant to the models. These terms were A, C, D, AB, AC, AD and CD, which had a significant effect to the MOE of the composite. As for the fit statistical model, the R^2 and adjusted R^2 value was relatively higher than the MOR model. This shows that the amount of variations around the mean were explained in the model created. However, there was a big difference between the adjusted R^2 value, and the predicted R^2 value, which means that the MOE response predicted for different observation may not be similar to the experimental response. The following was the generated model for MOE:

$$\begin{aligned} MOE = & -1.01351(Wood) + 0.241318(bamboo) + 0.064865(MMT) - 0.001091(PLA) \\ & + 0.009488(wood)(bamboo) + 0.010336(wood)(MMT) \\ & + 0.011422(wood)(PLA) - 0.001988(bamboo)(MMT) \\ & - 0.003004(bamboo)(PLA) - 0.000873(MMT)(PLA) \end{aligned} \quad (2)$$

By using the model, the following predicted MOE values were generated, as shown in Table 9. Figure 2 shows the plotted predicted graph against experimental graph, which was obtained from the Design Expert software.

Based on the graph in Fig. 2, the values were clearly off and slightly differed from the actual values of MOE. Furthermore, the predicted MOE value for Sample 5 was negative, which was due to the same reason, whereas a low R^2, adjusted R^2 and predicted R^2 were obtained for MOR model. The faulty composite was due to the uneven mixing or nonproper dispersion. In addition, as heat involved, this also affected the sample strength. Based on to Hagen [34], thermoplastic polylactic acid (PLA) melted at temperature 160–180°C and had a glass transition temperature at

Table 9 Comparison between the experimental and predicted results

Run	Component 1 A: Wood (wt%)	Component 2 B: Bamboo (wt%)	Component 3 C: MMT (wt%)	Component 4 D: PLA (wt%)	Experimental MOE (MPa)	Predicted MOE (MPa)
1	7	4.5	3.187	85.313	0.0303061	0.036448482
2	4.5	4.5	6.85	84.15	0.0107905	0.011577643
3	7	7	10	76	0.0257405	0.024558
4	5.52876	7	4.9796	82.4916	0.0100307	0.017301369
5	4.5	6.93	3.3503	85.2197	0.0156509	−0.001686942
6	7	4.5	7.42459	81.0754	0.0183703	0.014960847
7	4.5	6.05415	1	88.4459	0.0110177	0.021796679
8	4.5	4.5	9.17165	81.8284	0.0115483	0.012008356
9	7	4.5	10	78.5	0.0128882	0.0172205
10	5.69862	5.65713	9.64977	78.9945	0.0312569	0.028856435
11	6.02857	4.5	1	88.4714	0.0771753	0.073819479
12	5.69862	5.65713	9.64977	78.9945	0.0312569	0.028856435
13	4.5	7	10	78.5	0.0116233	0.015088
14	6.96394	5.55255	4.89852	82.585	0.0136255	0.013260717
15	5.69862	5.65713	9.64977	78.9945	0.0312569	0.028856435
16	7	7	1	85	0.0258964	0.024639
17	5.52876	7	4.9796	82.4916	0.0100307	0.017301369
18	7	7	6.99281	79.0072	0.0100802	15.2972351
19	6.96394	5.55255	4.89852	82.585	0.0136255	14.2796271
20	6.02857	4.5	1	88.4714	0.0771753	12.4338858

55–56°C. Longer time spend on the heated plate could resulted in the degradation other ingredients in the composites such as wood, bamboo and MMT, which affecting the strength of the composite. Another factor that affected the composite strength was presence of air bubbles, pores or void in the composite. The hot-press method squeezed the air, while closing the gap between the mold and the heated plate. Air bubbles were trapped in the composite as resultant due to compression process. The presence of air bubbles affected the composite cross-sectional area, whereas the tensile stress applied.

3.1.3 Optimization for Modulus of Rupture (MOR)

For the composite optimization, few criteria were defined, which involves the weight percentage of wood, bamboo, MMT and PLA, and the composite strength in terms of MOR. For this purpose, wood and bamboo were maximized, and MMT and PLA were minimized, to obtain the maximized value of MOR. The reason for maximizing wood

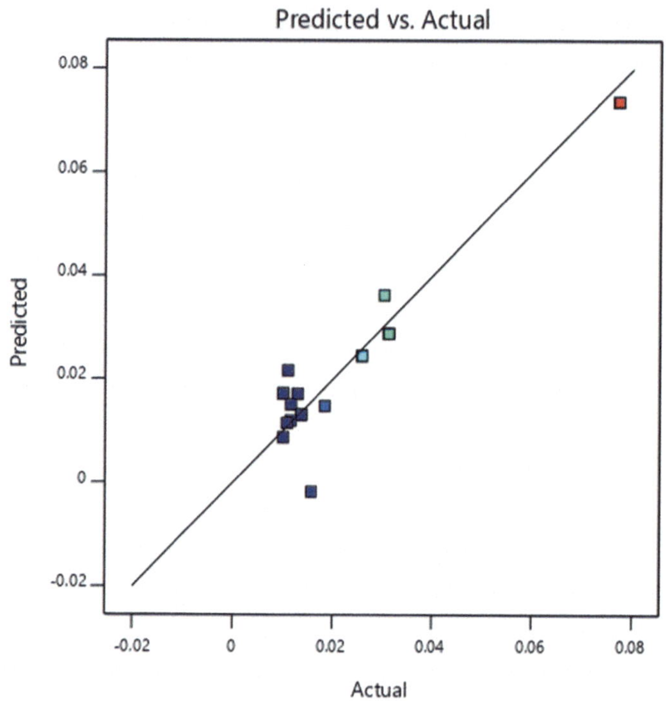

Fig. 2 Predicted vs actual graph for MOE

and bamboo composition was to utilize natural resources that was biodegradable, whereas it does not cause plastic pollution. This led to the utilization of ecofriendly material, which used to reduce plastic waste and greenhouse gas pollution that comes from the manufacturing of synthetic plastic. As for the MMT, the thermal degradation property was high, which lead to a strong composite that withstand high temperature condition. According to Leszcznska et al. [35], no significant mass loss happened due to the decomposition process at 300°C for pure montmorillonite. On the other hand, PLA was the main polymer set to be in the specified range as shown in Table 1. Finally, the composite strength was preferably to be maximized, whereas the resulting optimization was shown in Fig. 3.

Figure 3 shows the best ingredients composition was 7wt% wood, 7wt% bamboo, 10wt% MMT and 76wt% PLA, with resulting MOR of 16.3752 GPa. Therefore, the predicted value for Sample 3 had the optimized ingredients composition. In addition, Sample 3 also had the highest experimental MOR value, which was 15.8008 GPa and further proves the optimization generated by the Design Expert software.

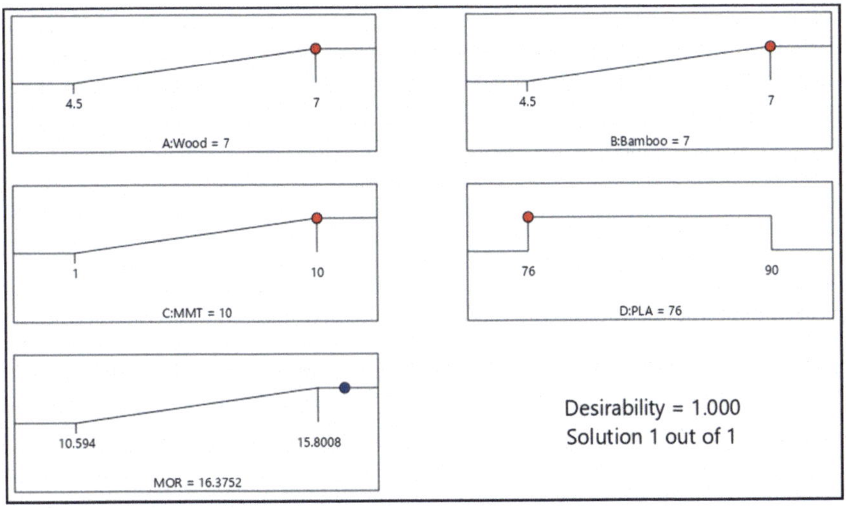

Fig. 3 Optimization for wood, bamboo, MMT, and PLA weight percentage for maximum composite MOR

3.1.4 Optimization for Modulus of Elasticity (MOR)

For the composite optimization, wood and bamboo usage was maximized, while maximum MOE value is obtained. By using the same criteria, the following optimization was created which was shown in Fig. 4.

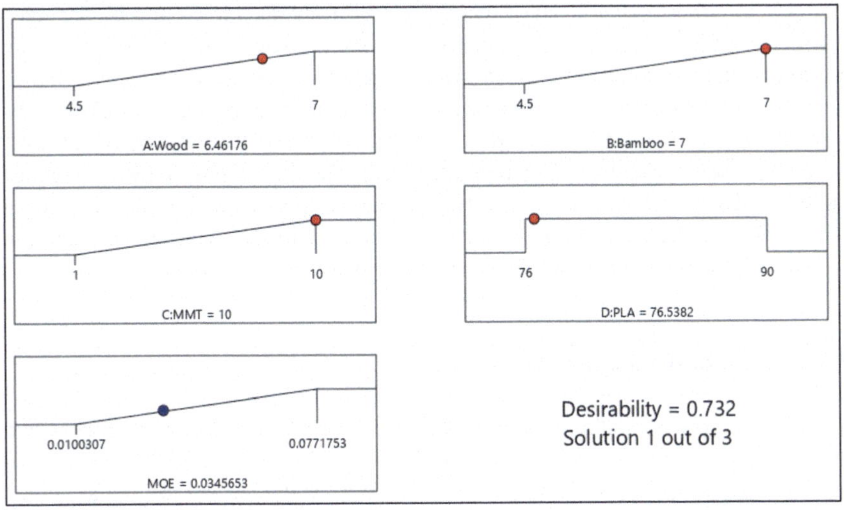

Fig. 4 Optimization for wood, bamboo, MMT and PLA weight percentage for maximum composite MOE

Based on the result, the optimized weight percentage of wood, bamboo, MMT and PLA was 6.46176, 7, 10 and 76.5382, respectively. This was produced a composite with MOE value of 0.0345653 GPa. This predicted MOE value was less accurate, since the model for MOE has a very low predicted R^2 value of 0.2884, which had a average error percentage of 73.56%.

3.2 *Morphological Analysis*

For the SEM analysis, several samples were used ranging from the preparation of the raw materials, which is the raw wood and bamboo until to the composite samples. For the composite sample, SEM was done on the rupture surface of the composite to analyze the effect of mechanical stress on the structure of the composite.

3.2.1 SEM of the Wood Fiber

Thus, SEM was carried out on the wood sample from before dewaxing as shown in Fig. 5(a), after dewaxing as shown in Fig. 5(b), after mercerization as shown in Fig. 5(c), after bleaching as shown in Fig. 5(d), and after ball milling Fig. 5(e).

Based on the figures, the significant changes were observed in Fig. 5(c) and Fig. 5(e). In Fig. 5(c), the number of smaller sized wood fiber was observed to be lower than Fig. 5(b). This was due to the mercerization process, whereas the fiber swelled during the post-treatment, while almost 25% of the hydrogen bonds broken [36]. It also noticed that the broken bonds started to re-bond with each other and this indicates the smaller sized of wood fiber has re-bonded to from a bigger sized particle. For Fig. 5(e), the changes in size was observed due to the wood fiber being ground by the ball milling machine, which made the particle size smaller. Thus, a higher magnification was used to observe the small particle due to ball milling. The wood fiber sizes varied from 10.5 μm to 34.0 μm, a lot smaller than before ground, which had sizes varied from 63.6 μm to 159 μm. The ground purpose was to reduce the wood size, which cause higher surface area that improve the interfacial bonding with MMT and PLA. According to Djebara et al. [37], the particles high aspect ratio mixed with the polymer resulted in high strength composite.

3.2.2 SEM of the Bamboo Fiber

The SEM images of bamboo fiber sample were observed before dewaxing as shown in Fig. 6(a), after dewaxing as shown in Fig. 6(b), after delignification as shown in Fig. 6(c), after mercerization as shown in Fig. 6(d), and after ball milling as shown in Fig. 6(e).

Similar with the wood fiber, significant changes were found after the sample undergoes mercerization and ball milling. For bamboo fiber after mercerization process

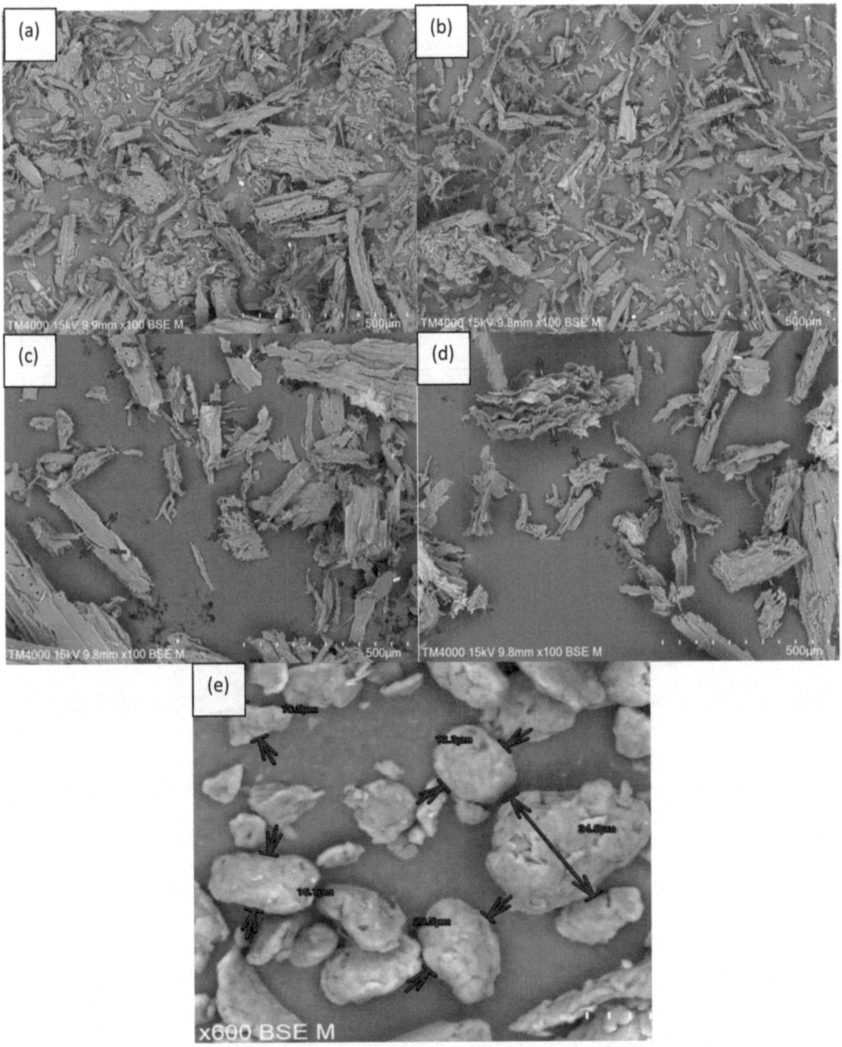

Fig. 5 Wood fiber (a) before dewaxing at × 100, (b) after dewaxing at × 100, (c) after mercerization at × 100, (d) after bleaching at × 100, and (e) after ball milling at × 600

was shown in Fig. 6(d), whereas the wood fibers were separated, strand and relatively longer. According to Khalil et al. [38], alkaline treatment or mercerization had several effects on bamboo fiber by increasing its surface roughness, decreasing the fiber diameter, and removal of lignin, fat, protein and non-water-soluble ingredients. Figure 6(d) show the bamboo fiber separation due to the removal of lignin that underwent delignification process. Mercerization has also reduced the bamboo fiber diameter. From the figure, the measured bamboo fibers diameter varied from 13.3 μm to 25.7 μm, while in Fig. 6(c), the fiber diameter varied from 31.1 μm to

Fig. 6 Bamboo fiber (a) before dewaxing at × 100, (b) after dewaxing at × 100, (c) after delignification at × 100, (d) after mercerization at × 100, and (e) after ball milling at × 600

283 μm. This large difference means that the alkaline treatment of the fibers was successful. Next significant change was seen from Fig. 6(e), whereas the long fibers form small particle. The small size particle was due to the ground process of ball milling that crushed the bamboo fiber into small particle varied in size from 5.27 μm to 40.8 μm.

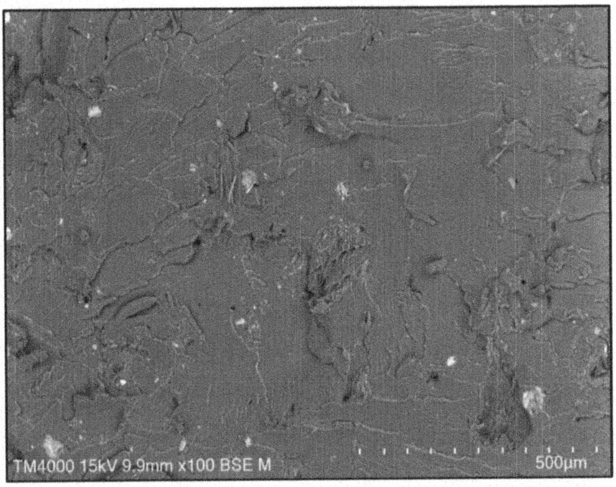

Fig. 7 SEM image for sample 20 composite at × 100

3.2.3 SEM of the Nanocomposite

Figure 7 shows the SEM image for Sample 20. From the figure, several white spots were observed, which was determine as bamboo and wood fibers particles. These fibers dispersed evenly throughout the composite and formed covalent bond with the PLA and MMT [33]. In addition, Sample 20 had the highest weight percentage of PLA, which resulted in the flat and smooth surface at the composite ruptured area. Air bubble had affected the mechanical properties of the composite, which resulted due to the produced design model become inaccurate as it comes to the predicting different observation. Figure 8 shows the samples that had air bubbles at the area of the mechanical testing rupture.

Based on the figure also, several large air bubbles that was due to the error during the hot press machine compression process. As result, the composites cross-sectional area had become inconsistent and affecting the mechanical strength of the composite. In Fig. 8(f), the largest air bubble was approximately around 300 μm in diameter.

3.3 Energy Dispersive X-Ray Analysis (EDX/EDS)

Based on Yesilkir-Baydar et al. [39], EDX was carried to analyze the elements of surface materials, which includes the amount of the elements. Table 10 shows the EDX of bamboo fiber particle. Based on the result, there were four elements detected, which is C, O, Al and Ag. The C and O element were detected for bamboo fiber that made up of cellulose that contains C-O bond [40]. However, there were traces of Al and Ag atom detected which has the mass percentage of 0.74 and 0.04, respectively a shown in Table 10. This was very low and concluded that the presence of Al and Ag was due to the impurities from the ball milling process.

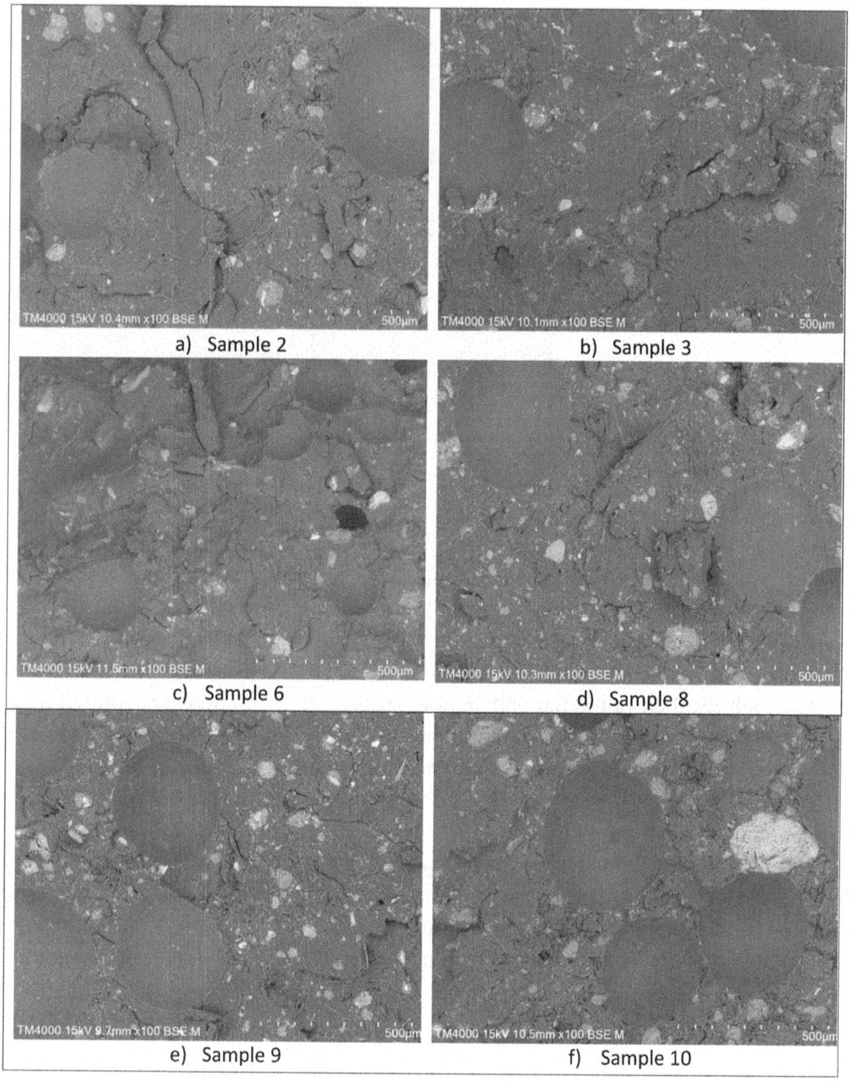

Fig. 8 Presence of air bubbles in some of the samples at × 100

As for wood fiber, the result was shown in Table 11. Based on the result, C and O elements was detected along with a trace of Al. Similar to the bamboo fiber EDX analysis, the C and O elements are attributed to the cellulose content, while the Al was due to impurities, as the mass percentage was very low. For the nanocomposites, Sample 18 was used, and the results was shown in Table 12. Based on the result, there were traces of Si and Ni elements contained in the composite, apart from C and O atom. According to Bergaya and Lagaly [41], an inorganic montmorillonite mineral was composed of aluminosilicate layers that had hydrated exchangeable cations in

Table 10 Summary generated by EDX analysis software for bamboo fiber

Element	At. No.	Netto	Mass (%)	Mass Norm. (%)	Atom (%)	Abs. error (%) (1 sigma)	Rel. error (%) (1 sigma)
C	6	105768	58.99	58.99	65.89	6.65	11.27
O	8	43853	40.23	40.23	33.73	4.82	11.97
Al	13	5281	0.74	0.74	0.37	0.06	8.42
Ag	47	165	0.04	0.04	0.00	0.00	9.16
		Sum	100.00	100.00	100.00		

Table 11 Summary generated by EDX analysis software for wood fiber

Element	At. No.	Netto	Mass (%)	Mass Norm. (%)	Atom (%)	Abs. error (%) (1 sigma)	Rel. error (%) (1 sigma)
C	6	90598	58.53	58.53	65.46	6.66	11.37
O	8	38975	40.65	40.65	34.13	4.91	11.09
Al	13	5220	0.82	0.82	0.41	0.07	8.10
		Sum	100.00	100.00	100.00		

Table 12 Summary generated by EDX analysis software for nanocomposite

Element	At. No.	Netto	Mass (%)	Mass Norm. (%)	Atom (%)	Abs. error (%) (1 sigma)	Rel. error (%) (1 sigma)
C	6	30802	52.80	52.80	60.25	6.51	12.33
O	8	19387	45.71	45.71	39.16	5.91	12.93
Si	14	2447	0.97	0.97	0.47	0.07	7.56
Ni	28	235	0.53	0.53	0.12	0.07	12.62
		Sum	100.00	100.00	100.00		

the spaces of the interlayer. Al-OH octahedral sheets in the aluminosilicate layer that was sandwiched by two Si-O tetrahedral sheets [41]. Thus, the Si element detected in the composite was due to montmorillonite, which had become the nano binder matrix.

3.4 *Fourier Transform Infrared Spectroscopy (FTIR)*

Sample 2 FTIR spectra was shown in Fig. 9 with Table 13. Based on Fig. 9, several peaks were observed in the wavenumber range of 3300–2900 cm^{-1}. There were 3739.97 cm^{-1}, 3232.7 cm^{-1} and 2926.01 cm^{-1} peaks. Based on Hospodarova et al.

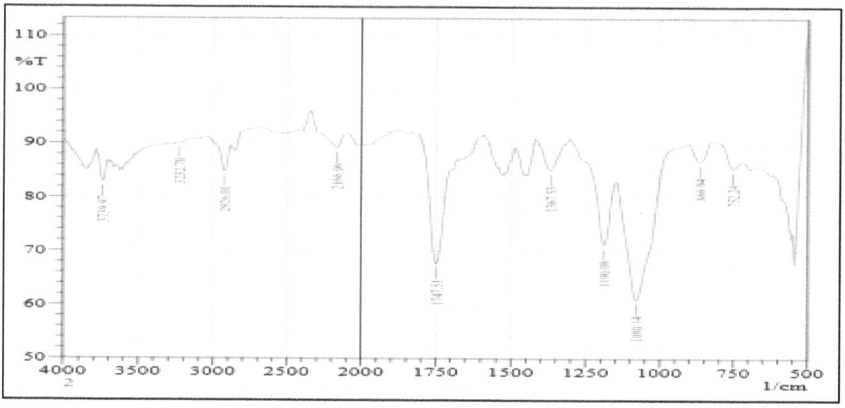

Fig. 9 IR spectra of composite Sample 2

Table 13 Summary of FTIR spectra for Sample 2

No.	Peak	Intensity	Corr. Inte	Base (H)	Base (L)	Area	Corr. Are
1	752.24	85.43	2.52	825.53	723.31	5.64	0.42
2	866.04	86.58	3.79	904.61	827.46	4.14	0.75
3	1080.14	60.94	24.19	1149.57	906.54	27.36	12.29
4	1190.08	71.2	13.92	1303.88	1151.5	12.96	3.69
5	1367.53	85.07	5.94	1419.61	1305.81	6.24	1.58
6	1747.51	67.6	20.59	1809.23	1699.29	11.21	5.36
7	2166.06	89.15	3.87	2345.44	2100.48	9.57	2.97
8	2926.01	84.79	5.29	3037.89	2885.51	8.15	1.49
9	3232.7	90.02	0.09	3242.34	3143.97	4.31	0.02
10	3739.97	82.79	5.01	3784.34	4705.26	5.46	1.03

[42], the range of 3660–2800 cm^{-1} peak was attributed to the C-H stretching vibration and O-H bonds in polysaccharides, which was the wood and bamboo fiber cellulose. On Fig. 9, the 3232.7 cm^{-1} peak had a broad characteristic. According to Hospodarova et al. [42], the broad peak was related to the hydroxyl group stretching vibration in polysaccharides, which includes the inter- and intra- molecular hydrogen bond vibration in the cellulose. Band at 2926.01 cm^{-1} was attributed to the C-H stretching vibration for all the hydrocarbon constituent in polysaccharides [42]. The presence of nanoclay on the other hand was observed at 3739.97 cm^{-1} peak band. The 1630–800 cm^{-1} absorption bands peaks were typically assigned to the cellulose content. Figure 9, the 866 cm^{-1}, 1080.14 cm^{-1}, 1190.08 cm^{-1} and 1367.53 cm^{-1} peaks was corresponded to the -CH and -CH_2C-O and -OH bending and stretching vibrations bonds in cellulose [40]. At 1747.51 cm^{-1} peak, carboxyl or carbonyl groups was found. Based on Wu et al. [43], the carboxyl or carbonyl groups was

assigned for intensity at 1700 cm^{-1}. For Si-O-Si inter tetrahedral bonds in SiO_2, the peak was identified at band 752.24 cm^{-1}. Wu et al. [43] stated that for untreated montmorillonite was in the range of 700–900 cm^{-1} bands, which includes C-O plane bending of carbonates, Si-O stretching of quartz, Al–Mg-OH bending, C-O in plane bending of carbonates and Al–Fe-OH bending.

In Fig. 10, Sample 20 FTIR spectra was distinct as compared to the of Sample 2 FTIR spectra. Based on the figure, there were only 4 peaks absorbed as summarized in Table 14, i.e. 1039.63 cm^{-1}, 1078.21 cm^{-1}, 1180 cm^{-1} and 1747.51 cm^{-1}. The peaks in the range of 1630-800 cm^{-1} represents the cellulose content characteristic. The peaks at 1039.63 cm^{-1}, 1078.21 cm^{-1} and 1180 cm^{-1} corresponded to the to the -CH and $-CH_2$, C-O and -OH bending and stretching vibrations bonds in cellulose. While, the 1747.5 cm^{-1} peak band was contributed to the carboxyl and carbonyl groups.

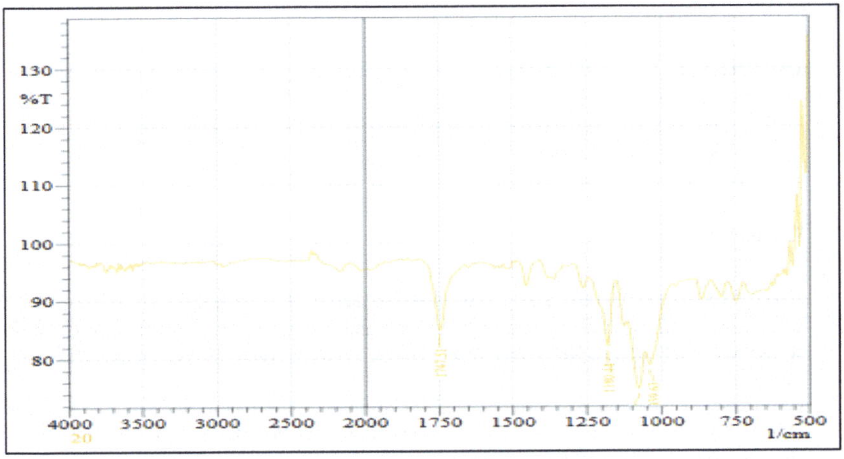

Fig. 10 IR spectra of composite Sample 20

Table 14 Summary of FTIR spectra for Sample 20

No.	Peak	Intensity	Corr. Inte	Base (H)	Base (L)	Area	Corr. Are
1	1039.63	78.903	4.31	966.34	966.34	6.097	0.785
2	1078.21	74.775	8.289	1056.90	1056.99	5.856	1.365
3	1180.44	82.11	11.497	1153.43	1153.43	4.444	1.888
4	1741.51	84.924	10.571	1813.09	1707	3.893	1.926

4 Conclusions

In conclusion, the strength of the nanocomposite was improved correspond to the increase in wood, bamboo and MMT weight percentage. In addition, the optimization agreed that the highest tensile strength was achieved at the highest weight percentage of 7wt% wood, 7wt% bamboo and 10wt% MMT. The strength of the nanocomposite with this material compositions was 15.8008 MPa for the experimental value, and 16.3752 MPa for the predicted value of the produced model. For the SEM analysis, mercerization and milling of the surface treatment of wood and bamboo fiber had a significant effect on the structure of the fiber. This structure was responsible for the interfacial reaction with the binder matrix such as MMT and PLA. During the mercerization, the lignin was removed from both wood and bamboo fiber which had cause the hydrogen bond to break and re-bonded again upon drying. Due to this, the diameter of the fibers was reduced and for the bamboo fiber, the length has increased due to the re-bonding.

Acknowledgements The authors would like to acknowledge Universiti Malaysia Sarawak (UNIMAS) for the support.

References

1. Dicker, M.P.M., Duckworth, P.F., Baker, A.B., Francois, G., Hazzard, M.K., Weaver, P.M.: Green composites: a review of material attributes and complementary applications. Compos. Part A: Appl. Sci. Manuf. **56**(1), 280–289 (2014). https://doi.org/10.1016/j.compositesa.2013.10.014
2. Akadiri, P.O., Chinyio, E.A., Olomolaiye, P.O.: Design of a sustainable building: a conceptual framework for implementing sustainability in the building sector. Build. **2**(2), 126–152 (2012). https://doi.org/10.3390/buildings2020126
3. Arora, A.K.: Environmental sustainability—necessary for survival. Environ. Sustain. **1**(1), 1–2 (2018). https://doi.org/10.1007/s42398-018-0013-3
4. Girijappa, Y.G.T., Rangappa, S.M., Parameswaranpillai, J., Siengchin, S.: Natural fibers as sustainable and renewable resource for development of eco-friendly composites: a comprehensive review. Front. Mater.: Polym. Compos. Mater. **6**(1), 226 (2019). https://doi.org/10.3389/fmats.2019.00226
5. Abdul Khalil, H.P.S., Bhat, I.U.H., Jawaid, M., Zaidon, A., Hermawan, D., Hadi, Y.S.: Bamboo fibre reinforced biocomposites: a review. Mater. Des. **42**(1), 353–368 (2012). https://doi.org/10.1016/j.matdes.2012.06.015
6. Abdul Khalil, H.P.S., Alwani, M.S., Islam, M.N., Suhaily, S.S., Dungani, R., H'ng, Y.M., Jawaid, M.: The use of bamboo fibres as reinforcements in composites. In: Faruk, O., Sain, M. (eds.) Biofiber Reinforcements in Composite Materials, pp. 488–524. Woodhead Publishing, Cambridge (2015). https://doi.org/10.1533/9781782421276.4.488
7. Okubo, K., Fujii, T., Yamamoto, Y.: Development of bamboo-based polymer composites and their mechanical properties. Compos. Part A: Appl. Sci. Manuf. **35**(3), 377–383 (2004). https://doi.org/10.1016/j.compositesa.2003.09.017
8. Mohammed, L., Ansari, M.N.M., Pua, G., Jawaid, M., Islam, M.S.: A review on natural fiber reinforced polymer composite and its applications. Int. J. Polym. Sci. 1, 1–15 (2015). https://doi.org/10.1155/2015/243947

9. Shen, L., Yang, J., Zhang, R., Shao, C., Song, X.: The benefits and barriers for promoting bamboo as a green building material in China—an integrative analysis. Sustain. **11**(9), 2493 (2019). https://doi.org/10.3390/su11092493
10. Atanda, J.: Environmental impacts of bamboo as a substitute constructional material in Nigeria. Case Stud. Constr. Mater. **3**(1), 33–39 (2015). https://doi.org/10.1016/j.cscm.2015.06.002
11. Rajak, D.K., Pagar, D.D., Menezes, P.L., Linul, E.: Fiber-reinforced polymer composites: manufacturing, properties, and applications. Polym. **11**(10), 1667 (2019). https://doi.org/10.3390/polym11101667
12. Song, X., Peng, C., Ciais, P., Li, Q., Xiang, W., Xiao, W., Zhou, G., Deng, L.: Nitrogen addition increased CO2 uptake more than non-CO2 greenhouse gases emissions in a Moso bamboo forest. Sci. Adv. (2020). https://doi.org/10.1126/sciadv.aaw5790
13. Silva, J.M.N., Carreriras, J.M.B., Rosa, I., Pereira, J.M.C.: Greenhouse gas emissions from shifting cultivation in the tropics, including uncertainty and sensitivity analysis. J. Geophys. Res.: Atmos. **116**(D20), 1–21 (2011). https://doi.org/10.1029/2011JD016056
14. Mohamed, A., Appanah, S.: Bamboo resources conservation and utilization in Malaysia. In: Rao, A.N., Rao, V.R. (eds.) Bamboo—Conservation, Diversity, Ecogeography, Germplasm, Resource Utilization and Taxonomy (1998). https://www.bioversityinternational.org/fileadmin/bioversity/publications/Web_version/572/ch27.htm
15. Youssefian, S., Rahbar, N.: Molecular origin of strength and stiffness in bamboo fibrils. Sci. Rep. **5**(1), 11116 (2015). https://doi.org/10.1038/srep11116
16. Peng, Y., Nair, S.S., Chen, H., Yan, N., Cao, J.: Effects of lignin content on mechanical and thermal properties of polypropylene composites reinforced with micro particles of spray dried cellulose nanofibrils. ACS Sustain. Chem. & Eng. **6**(8), 11078–11086 (2018). https://doi.org/10.1021/acssuschemeng.8b02544
17. Yang, J., Ching, Y.C., Chuah, C.H.: Applications of lignocellulosic fibers and lignin in bioplastics: a review. Polym. **11**(5), 751 (2019). https://doi.org/10.3390/polym11050751
18. Sharma, B., Shah, D.U., Beaugrand, J., Janecek, E.-R., Scherman, O.A., Ramage, M.H.: Chemical composition of processed bamboo for structural applications. Cellul. **25**(1), 3255–3266 (2018). https://doi.org/10.1007/s10570-018-1789-0
19. Lu, T., Jiang, M., Jiang, Z., Hui, D., Wang, Z., Zhou, Z.: Effect of surface modification of bamboo cellulose fibers on mechanical properties of cellulose/epoxy composites. Compos. Part B: Eng. **51**(1), 28–34 (2013). https://doi.org/10.1016/j.compositesb.2013.02.031
20. Liew, F.K., Hamdan, S., Rahman, M. R., Rusop, M. Thermomechanical properties of jute/bamboo cellulose composite and its hybrid composites: the effects of treatment and fiber loading. Adv. Mater. Sci. Eng. 1, 1–10 (2017). https://doi.org/10.1155/2017/8630749
21. Huang, Y. & Fei, B. Comparison of the mechanical charactersitics of fibers and cell walls from moso bamboo and wood. BioResource, 12(4), 82230–88239 (2017). https://doi.org/10.15376/biores.12.4.8230-8239
22. Chaowana, P.: Bamboo: an alternative raw material for wood and wood-based composites. J. Mater. Sci. Res. **2**(2), 90–102 (2013). https://doi.org/10.5539/jmsr.v2n2p90
23. Mohanty, A.K., Misra, M., Drzal, L.T.: Sustainable bio-composites from renewable resources: opportunities and challenges in the green materials world. J. Polym. Environ. **10**(1), 19–26 (2002). https://doi.org/10.1023/A:1021013921916
24. Hatami, M., Dehghan, A., Djafarzadeh, N.: Investigations of addition of low fractions of nanoclay/latex nanocomposite on mechanical and morphological properties of cementitious materials. Arab. J. Chem. **11**(6), 970–980 (2018). https://doi.org/10.1016/j.arabjc.2018.03.018
25. Bhattacharya, M.: Polymer nanocomposites—a comparison between carbon nanotubes, graphene, and clay as nanofillers. Mater. **9**(4), 262 (2016). https://doi.org/10.3390/ma9040262
26. Garusinghe, U.M., Varanasi, S., Raghuwanshi, V.S., Garnier, G., Batchelor, W.: Nanocellulose-montmorillonite composties of low water vapour permeability. Colloids Surf. A: Phys.Chemical Eng. Asp. **540**(1), 233–241 (2010). https://doi.org/10.1016/j.colsurfa.2018.01.010
27. Jintakosol, T., Nitayaphat, W.: Adsorption of silver (I) from aqueous solution using chitosan/montmorillonite composite beads. Mater. Res. **19**(5), 1114–1121 (2016). https://doi.org/10.1590/1980-5373-MR-2015-0738

28. ASTM D638-14: Standard test method for tensile properties of plastics, ASTM International, West Conshohocken, PA (2014). https://doi.org/10.1520/D0638-14
29. ASTM E2015-04: Standard guide for preparation of plastics and polymeric specimens for microstructural examination, ASTM International, West Conshohocken, PA (2014). https://doi.org/10.1520/E2015-04R14
30. ASTM E1508-12: Standard guide for quantitative analysis by energy-dispersive spectroscopy, ASTM International, West Conshohocken, PA (2019). https://doi.org/10.1520/E1508-12AR19
31. ASTM E168-16: Standard practices for general techniques of infrared quantitative analysis, ASTM International, West Conshohocken, PA (2016). https://doi.org/10.1520/E0168-16
32. ASTM E1252-98: Standard practice for general techniques for obtaining infrared spectra for qualitative analysis, ASTM International, West Conshohocken, PA (2013). https://doi.org/10.1520/E1252-98R13E01
33. Adamu, M., Rahman, M.R., Hamdan, S.: Formulation optimization and characterization of bamboo/polyvinyl alcohol/clay nanocomposite by response surface methodology. Compos. Part B: EngineeringComposite Part B: Eng. **176**(1), 107297 (2019). https://doi.org/10.1016/j.compositesb.2019.107297
34. Hagen, R.: Polylactic acid. Polym. Sci.: Compr. Ref. **10**(1), 231–236 (2012). https://doi.org/10.1016/B978-0-444-53349-4.00269-7
35. Leszczynska, A., Njuguna, J., Pielichowski, K., Banerjee, J.R.: Polymer/montmorillonite nanocomposites with improved thermal properties. Part I. Factors influencing thermal stability and mechanisms of thermal stability improvement. Thermochim. Acta 453(1), 75–96 (2007). https://doi.org/10.1016/j.tca.2006.11.003
36. Dai, D., Fan, M.: Wood fibres as reinforcements in natural fibre composites: structure, properties, processing and applications. In: Hodzic, A., Shanks, R. (eds.) Natural Fibre Composites: materials, Processes and Properties, pp. 3–65. Woodhead Publishing, Cambridge (2014). https://doi.org/10.1533/9780857099228.1.3
37. Djebara, Y., Moumen, E.L., Kanit, T., Madani, S., Imad, A.: Modeling of the effect of particle size, particles distribution and particles number on mechanical properties of polymer-clay nano-composites: numerical homogenization versus experimental results. Compos. Part B: Eng. **86**(1), 135–142 (2016). https://doi.org/10.1016/j.compositesb.2015.09.034
38. Abdul Khalil, H.P.S., Salwani, M., Islam, M. N., Suhaily, S.S., Dungani, R., Jawaid, M.: The use of bamboo fibres as reinforcements in composites. In: Faruk, O., Sain, M. (eds.) Biofiber Reinforcements in Composite Materials, pp. 488–524. Woodhead Publishing, Cambridge (2015). https://doi.org/10.1016/C2013-0-16470-7
39. Yesilkir-Baydar, S., Oztel, O.N, Cakir-Koc, R., Candayan, A.: Evaluation techniques. In: Razavi, M., Thakor, A. (eds.) Nanobiomaterials Science, Development and Evaluation, pp. 211–232, Woodhead Publishing, Cambridge (2017). https://doi.org/10.1016/B978-0-08-100963-5.00011-2
40. Xu, F., Yu, J., Tesso, T., Dowell, F., Wang, D.: Qualitative and quantitative analysis of lignocellulosic biomass using infrared techniques: a mini-review. Appl. Energy **104**(1), 801–809 (2013). https://doi.org/10.1016/j.apenergy.2012.12.019
41. Bergaya, F., Lagaly, G.: General introduction: clays, clay minerals, and clay science. Dev. Clay Sci. **1**(1), 1–18 (2013). https://doi.org/10.1016/S1572-4352(05)01001-9
42. Hospodarova, V., Singovszka, E., Stevulova, N.: Characterization of cellulosic fibers by FTIR spectroscopy for their further implementation to building materials. Am. J. Anal. Chem. **9**(6), 303–310 (2018). https://doi.org/10.4236/ajac.2018.96023
43. Wu, L.M., Tong, D.S., Zhao, L.Z., Yu, W.H., Zhou, C.H.: Fourier transform infrared spectroscopy analysis for hydrothermal transformation of microcrystalline cellulose on montmorillonite. Appl. Clay Sci. **95**(1), 74–82 (2014). https://doi.org/10.1016/j.clay.2014.03.014

Bamboo Cellulose Gel/MMT Polymer Nanocomposites for High Strength Materials

Md Rezaur Rahman and Muhammad Khusairy Bin Bakri

Abstract This research focuses on the interrelationship between the structure and properties of cellulosic composites. To achieve its optimum performance, the compatibility of produced nanocomposites was focus. Few studies had shown that the properties of biopolymers enhanced by the presence of nanofillers. This was proved from the characterization analyses, involving structural, thermal and mechanical properties of the nanocomposites using the Fourier transform infrared (FTIR) spectroscopy, scanning electron microscopy (SEM), energy dispersive x-ray (EDS/EDX), Brunauer-Emmett-Teller (BET), tensile strength and tensile modulus tests. For the mechanical properties, the highest tensile strength and tensile modulus was achieved for Sample 10 which contain 4wt% cellulose, 2.434wt% MMT and 93.566wt% of PLA. The strength generated by this composite was 16.063 MPa for the tensile strength and 0.015083 GPa for the tensile modulus.

Keywords Bamboo · MMT · Cellulose · Nanocomposites · Gel

1 Introduction

In the age of digitalization and vast development, conserving and protecting the environment had become a main topic [1]. This had led to the exploration of new alternative by many researchers, especially those sustainable materials, which is renewable in the resource [1, 2]. One of the alternative approaches in terms of material design was using a raw materials renewable source [2]. Moreover, many researchers had found functionality of incorporated bamboo materials mostly related to its material design [3]. Bamboo was one of grasses family that required less regeneration growth, especially in those develop stems throughout asexual reproduction without replanting [4]. Bamboo was also found to abundantly exist in Asia and South America and was used for building facilities and equipment [5]. Moreover, in Malaysia the bamboo growth was distributed evenly. Based on Ng and Noor [6], Malaysian bamboo was

M. R. Rahman (✉) · M. K. B. Bakri
Faculty of Engineering, Universiti Malaysia Sarawak, Jalan Datuk Mohammad Musa, 94300, Kota Samarahan, Sarawak, Malaysia
e-mail: rmrezaur@unimas.my

M. R. Rahman (ed.), *Bamboo Polymer Nanocomposites*, Engineering Materials,
https://doi.org/10.1007/978-3-030-68090-9_7

typically found in localized riverbank patches, lowland forest, hillsides, and ridge tops. The bamboo population was also mixed with other species.

Based on Kamaruzaman [7], approximately 587 million bamboo culms in Malaysian forest. Wong [8] mentioned that about 70 species of bamboo owns by Malaysia covered about 50% from West Malaysia, 30% from Sabah and 20% from Sarawak. The common species were *Bambusa, Dendrocalamus, Chusquea, Dinochloa, Phyllostachys, Thyrsostachys, Yushania, Schizostachyum, Gigantochloa,* and *Racemobambos* and most widespread species were *Gigantochloa, Dendrocalamus* and *Schizostachyum* [9]. Compared with synthetic polymer, Bamboo was known as a millennial crop, due to its excellent mechanical, physical and chemical characteristics. Based on Awoyera and Ugwu [5], bamboo had the potential like steel for construction materials due to its high strength compared to other materials such as brick, concrete, and wood. Roslan et al. [10] also claimed that bamboo was a genuinely environmentally friendly, cheap, non-abrasive, biodegradable, sustainable, and non-toxic plant. About 78% of cellulose was found in bamboo fiber, and the remainder comprises 12.5% hemicellulose, 10.1% lignin, 0.4% pectin and 3.2% aqueous extract [11].

Bamboo cellulose was an organic polymer. It was classified as polysaccharides consisting of β-1,4 linear chain linked with D-glucose unit, which had the polymerization degree range from hundreds up to ten thousand [12]. Several bamboo drawbacks were due to inclusion of its raw bamboo chemical components such as lignin, cellulose, wax and hemicellulose, which hindering the binding of bamboo and polymer matrix [10]. The drawbacks contributed to weak interfacial the fiber-matrix bonding that led to a decrease in mechanical properties. Cellulose and its derivatives were suitable for building gel blocks. Based on Menon et al. [13], cellulose gel was extracted from corn, bamboo, and sugarcane plants in many ways. Cellulose gel or microcrystalline cellulose (MCC) was made from natural polymer and depending on their characteristics, it was commonly used in the field of absorbents, medicine, biomaterials, and biosensors, [14]. Nevertheless, limited gel strength tends to restrict its application. Consequently, to address their weakness, low-cost inorganic functional fillers were introduced.

Clay minerals had created much attention among the researchers, especially as an additive in polymer reinforcement, due to its intercalation properties and smaller particle sizes. Montmorillonite (MMT) was widely used natural clay. Other than MMT, graphene and carbon nanotubes nanofillers were also shown promising, especially its property profiles to be used in polymer. Due to their relatively high cost and to date their usage has been reduced [15]. Based on Shetti et al. [16], MMT was part of the smectite clay group. The MMT particle structure had a typically persevered in sheets and layers, whereas each layer consists of two different octahedral and tetrahedral sheets [17]. Rahman et al. [18], showed that MMT lamellar elements had a high degree of stiffness, strength, and ratio in most organic applications. Uddin [19] also stated for composite materials, MMT was an excellent heat resistant and insulator. The natural inorganic nanostructure such as MMT had a good absorption behavior and high tolerance towards organic molecules. Hence, this work focused to improve the performance of nanocomposite cellulose gel, in addition of montmorillonite to overcome the barriers. The quality of nanocomposites was examined using several techniques and test equipment.

2 Methodology

2.1 Materials

The bamboo was obtained from Kampung Daun, Singai, Bau, Sarawak. Toluene, ethanol and montmorillonite MMT were obtained from Sigma Aldrich. Hydrogen peroxide was obtained from Qeric, acetic acid was obtained from Ensure, sodium hydroxide was obtained from Merck KgaA and titanium (IV) oxide was obtained from JT Baker. All the listed chemicals are under analytical grade.

2.2 Response Surface Methodology (RSM) by Design Expert

In research, an experimental design was a technique for determining factors that could affect the output results. To reduce the design costs, engineering design adjustments, product materials wastage and labor complexity by reducing process variability, this approach was used to achieve optimal efficiency. For optimization, response surface methodology (RSM) was a multivariate analytical technique that had a broad library of mathematics and statistical method. The RSM referred to the polynomial equation that describe to the behavior of the data set. For the preparation of bamboo cellulose/MMT nanocomposite, it explored the optimal experimental condition, to achieve the best possible results and a minimum number of experiments. The statistical software 'Design Expert® 11.0' from StatEase, Inc., was used to analyze the preliminary design. Table 1 shows the variables used to generate the material compositions for nanocomposite. Table 2 shows the generated 16 different nanocomposites materials compositions, whereas the tensile tests were carried out.

Table 1 Variables in the design expert

Component	Low level	High level
Bamboo cellulose (MCC)	4	8
Montmorillonite (MMT)	1	10
Polylactic Acid (PLA)	82	95

Table 2 Input variables for cellulose (MCC), PLA and MMT for preparation of polymer nanocomposite film

Designation sample	MCC (wt%)	MMT (wt%)	PLA (wt%)
0	0	0	100
1	5.71599	6.24423	88.0398
2	7.571597645	10	82.4284
3	6.410032664	1	92.58996734
4	8	4.080894568	87.91910543
5	4	2.433708879	93.56629112
6	7.571597645	10	82.42840236
7	6.410032664	1	92.58996734
8	8	1.776603371	90.22339663
9	4	10	86
10	4	2.433708879	93.56629112
11	4	5.530274037	90.46972596
12	8	6.25684681	85.74315319
13	5.725992639	6.244230501	88.03977686
14	5.715992639	6.244230501	88.03977686
15	5.844633171	3.422588998	90.73277783
16	6.363687197	8.588197699	85.0481151

2.3 *Bamboo Fibers Preparation*

Green bamboo culm was cut in 1 m length. The thin bamboo layer that includes the endodermis and exoderms bark was removed, as well as the node portions of the bamboo. Bamboo chips are produced by ground the bamboo culm using a grinder. The chips were dried in the oven for 96 h at 48.5°C. The dried sample was ground and sieved using 600 μm size sieve. The sieved sample was synthesized to obtain its cellulose.

2.4 *Cellulose Extraction from Bamboo Fibers*

2.4.1 Dewaxing Process

A toluene/ethanol solution with ratio of 2:1 was prepared by filled in a flask containing 400 ml toluene and 200 ml ethanol. The solution was heated and a Soxhlet extractor was placed on top of the boiling flask. To secure the Soxhlet extractor, a retort stand was used. Added 10 g of bamboo fibers was added into the membrane tube and placed into the extraction thimble. The Liebig condenser was placed on top of the extractor and attached firmly. To measure and control the temperature, a thermometer was used to maintain the sample at 250°C. The extraction process was continued for 6 h until the color disappeared in the specimen. After undergoing bleaching process, a tweezer was used to remove the extraction thimble. The sample was poured into a beaker and simultaneously stir with the addition of toluene-ethanol solution that was prepared previously. The sample was placed to cool. After cooling, the suspension was filtered using Whatman filter paper using filter funnel. The residue was labelled as dewaxed bamboo fiber (DBF) and placed overnight in the oven at 70°C.

2.4.2 Delignification Process

Delignification solution was prepared using 35wt% hydrogen peroxide weighed at 82.3 g and 99.8 wt% acetic acid weighed at 106.2 g. Titanium (IV) oxide was added to the solution and acts as the catalyst. 30 g of dry DBF sample was added into the delignification solution and the sample was heated for 2 h at 130°C. After 2 h, the heater was switched off, and the sample was left to cool at room temperature. The sample was filtered using Bunchner flask and rinsed using deionized water until exceeded pH 7 in order to prevent any generation of strong hydrogen bonds between the cellulose bundles. The sample was labeled as acid-treated cellulose fiber (ALCFs) and dried in the oven for 24 h at 70°C.

2.4.3 Mercerization Process

Mercerization process was carried out by adding ALCFs to 6 wt% of sodium hydroxides. The mixture was stirred and heated using auto shaker at 150 rpm for 2 h at temperature of 80°C. After heated for 2 h, the sample was stirred for 8 more h to remove hemicelluloses, pectin's and residual starch residual from the sample. The treated sample was filtered and rinsed using Bunchner flask and deionized water until reached the pH 7. The sample was freeze-dried at −85 °C for 48 h.

2.5 Nanoscale Cellulose Preparation Using Mechanical Fibrillation Process

As the sample dried, it was continued by ground it using grinder for 10 minutes and stop-dried for another 10 minutes. The step was repeated for several times until fine powder cellulose was obtained. The sample was labeled as cellulose fiber (CF). A mechanical technique was used to obtain nanoscale CF. Ball milling was used to convert the particles size of CF. The milling process was done at a speed of 600 rpm for 7 h using steel ball. The milling time, rotation speed and mass of the milling medium were kept constant in order to obtain the nano size properties of the sample.

2.6 *Nanocomposite Film Preparation*

By referring to the materials compositions for nanocomposite as shown in Table 2, the bamboo fiber, MMT and PLA were weighted and poured mix homogeneously. The mixture samples were hot pressed using hot-press machine at temperature 180°C for 15 minutes. After the compression, the mold was left to cool and cured, and the produced nanocomposite was removed from the mold.

2.7 *Characterization*

2.7.1 Fourier Transform Infrared Spectroscopy (FTIR)

FTIR 'IRAffinity-1' spectroscopy (Shimadzu; Japan) was used to identify the functional groups and the structures of molecular bond through the IR spectrum bands at a range of 4000 to 400 cm^{-1}. Analysis of qualitative and quantitative on the FTIR spectrum was carried out based ASTM E1252-98 [20] and ASTM E168-16 standards [21].

2.7.2 Scanning Electron Microscopy (SEM) and Energy Dispersive X-Ray (EDX/EDS)

Hitachi Analytical Tabletop SEM (benchtop) 'TM-3030' (Hitachi High-Technologies (Germany) Europe GmbH.) was used to take the morphological images of the samples. The sample was mounted on the stubs of aluminum and fine coated using 'JFC-1600' (JEOL (Japan) Ltd.). The images of the composites surface were collected using field emission gun with voltages of 5 kV and 15 kV. The test was performed based on ASTM E2015-04 [22] standard.

2.7.3 Brunauer-Emmett-Teller (BET)

The Brunauer-Emmett-Teller (BET) was conducted to study the specific surface area of 3rd and 4th bamboo cellulose cycle after underwent ball milling process using nitrogen gas adsorption method at 77 K using Autosorb iQ Any Gas (AG) Sorption. The specific surface area measurement was determined in the total surface area against the mass of the sample (m^2/g) unit. The BET characterization test was carried out with a degassing temperature of 200°C for 5 h. The average pore size of irregular porous samples is estimated.

2.8 Mechanical Properties

For tensile strength and tensile modulus test, the tensile tests were carried out using Universal Testing Machine (UTM) (T-machine Technology Machine (Taiwan) Co. Ltd.) at 5 mm/min crosshead speed based on ASTM D638-14 [23] standard. Extensometer was used to identify the tensile modulus and elongation.

3 Results and Discussions

3.1 Fourier Transform Infrared Spectrophotometer (FTIR) of Bamboo Cellulose

Figures 1, 2, 3, and 4 show the identification of absorption peaks for each chemical component present in the tested samples. Based on IR spectra, the peak intensity of 4000–2995 cm^{-1} was attributed to the presence variation of O-H stretching. A strong O-H stretching variation was observed in all FTIR results, at the peak intensity of 3331.07, 3286.70, 3336.85, and 3331.07 cm^{-1}, respectively. It was found out that the reformation of intermolecular hydrogen bonds between the free hydroxyl with the water molecules increases made the band become less intense [24]. This was due to a breakage of intermolecular and intramolecular hydrogen bonds in the cellulose main chain. Besides, the C-H band, it was observed at the peak intensity of 2891, 2885.51, 2897.08 and 2891.30 cm^{-1} in Figs. 1, 2, 3 and 4. Zheng et al. [24] characterized the

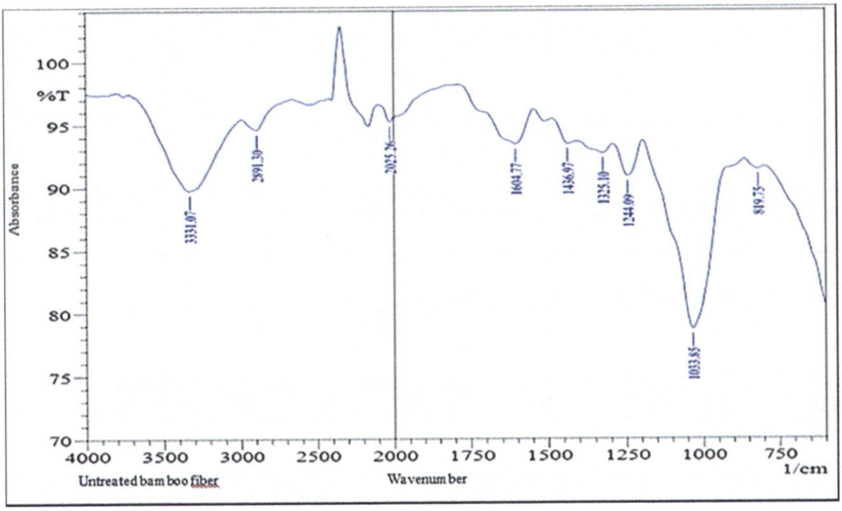

Fig. 1 IR spectra of untreated bamboo fiber

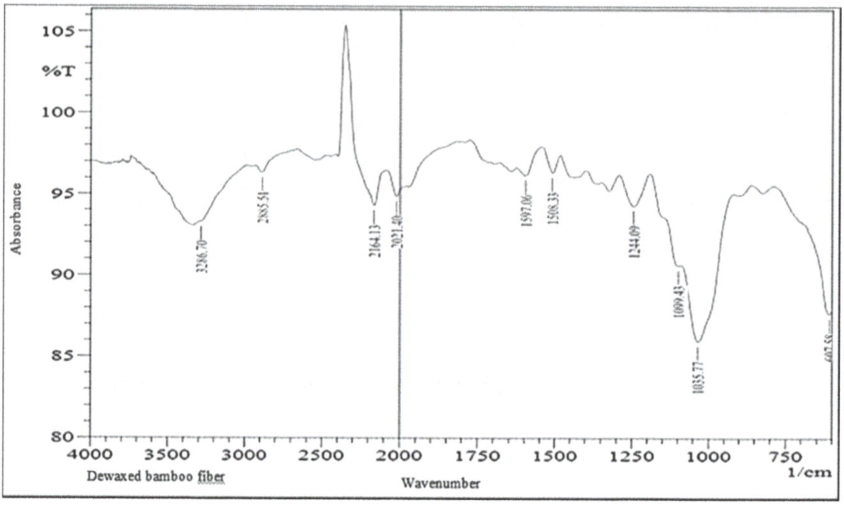

Fig. 2 IR spectra of dewaxed bamboo fiber

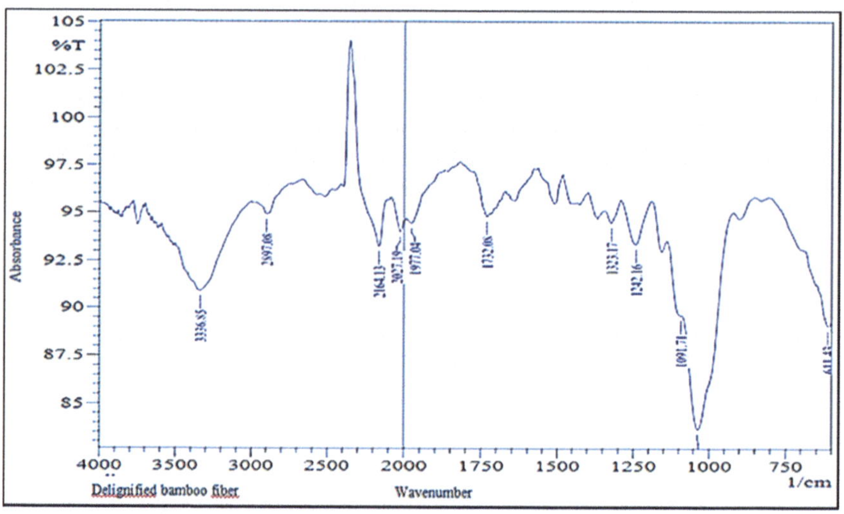

Fig. 3 IR spectra of delignified bamboo fiber

aliphatic moieties band as symmetric and asymmetric C-H groups in polysaccharides. In Fig. 1, the peak mainly at 1244.09 cm^{-1} shows the attribution of variation of aryl-alkyl ether C-O-C stretching. The C-H aromatic hydrogen linkage also present in the FTIR result of untreated bamboo fiber, which was contributed by 819.75 cm^{-1} peak intensity.

The lignin was identified between the ranges of 1500–1600 cm^{-1} that corresponds to aromatic skeletal vibration. At 1597.06 cm^{-1} peak in Fig. 2 was the dewaxed

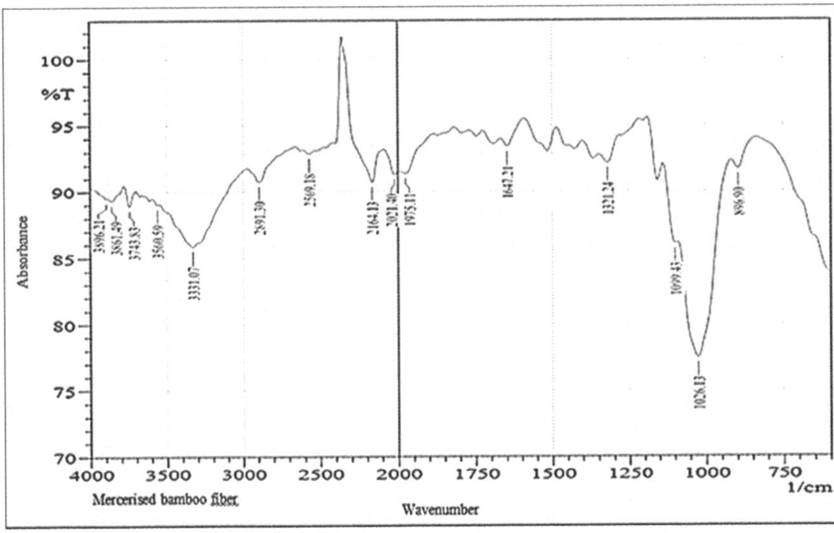

Fig. 4 IR spectra of mercerized bamboo fiber

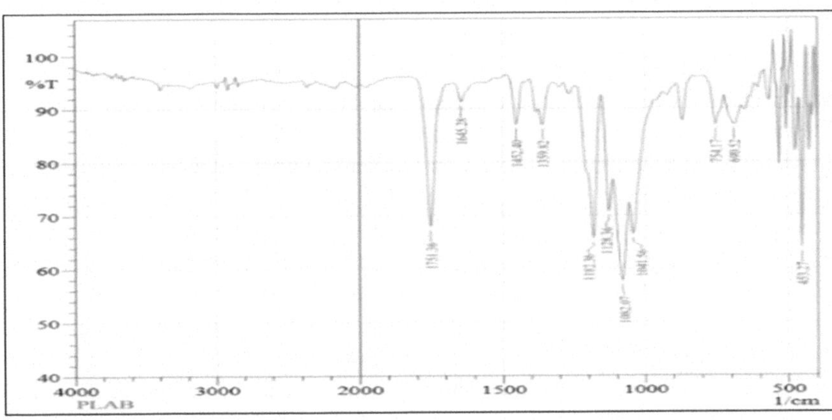

Fig. 5 IR spectra of pure PLA

bamboo cellulose, which refer to the C = C stretching of an aromatic skeletal functional group of lignin. However, the peak intensity in Figs. 3 and 4, due to C = C stretching was eliminated. Acid hydrolysis caused structural modification which had an impact on aryl-alkyl ether cleavage [25]. The lignin degradation involves the aryl ether bonding that initiated the phenolic components, particularly on the ferulic acid and p-coumaric acid. Nguyen et al. [26] mentioned that the peak intensity at 1732.08 cm^{-1} in delignified sample spectrum, either related to the presence of uronic ester and acetyl hemicellulose constituents or p-coumaric acids and carboxylic acid ferulic in lignin and hemicellulose. Gabhane et al. [27], and Li et al. [28] stated that the absorption spectrum mentioned was linked to the feruloyl, acetyl and p-coumaroyl groups of both lignin and hemicellulose. However, after the alkali treatment, the p-cuomaric rupture bond resulted in the elimination of hemicellulose and lignin.

In order to obtain a high content of cellulose, delignification process was performed using acetic acid, to remove non-cellulosic components. However, acetylation occurred as acetic acid was used for delignification acid hydrolysis of bamboo cellulose, which was identified with the presence of acetyl ester band ($C = O$) peak. The peak within the range of 1765–1715 cm^{-1} intensity expressed as the high amount of $C = O$ bond. In Fig. 3, the 1732.08 cm^{-1} peak intensity was observed and representing $C = O$ symmetrical linkage for ketone and carbonyl of acetyl group. The appearance of these peak had two possibilities: (i) Due to hemicellulose, which still present in the sample, even though it was treated with acid. The hemicellulose $C = O$ linkage was between 1765–1715 cm^{-1} bands, and (ii) Due to the aldehyde or carboxyl absorption development from the oxidizing C-OH groups. The peak intensity band at 1732.08 cm^{-1} was most likely assigned to oxidation as the color of cellulose after the delignification process is yellow during observation.

The 1325.10, 1436.97, 1323.17, and 1321.24 cm^{-1} peak intensities in Figs. 1, 3 and 4 show the medium C-H and $–CH_2$ bending characteristics. These bands ranges were specified as the crystallinity band, which shows the crystallinity degree in the sample, whereas higher rate was observed when these peaks are of high intensity [24]. The strong C-H aromatic and CH_2 intensity band was observed at the peak of 819.75 cm^{-1} in the raw sample, and 896.90 cm^{-1} in mercerized sample. The peak between 600–700 cm^{-1} indicated the intense C_1-H deformation which was identified in dewaxed cellulose at 607.58 cm^{-1} and delignified cellulose at 611.41 cm^{-1}.

In addition, the 896.90 cm^{-1} peak was correlated with the glycosidic C_1-H deformation and O-H bending group. Liew et al. [11] indicated that this peak was related to the β-glycosidic linkage in anhydoglucose units of cellulose. In addition, the 896.90 cm^{-1} band was corresponded to the asymmetrical stretching out of the plane. The peak at 1647.21 cm^{-1} in Fig. 4 expressed as H-O-H stretching variation. The 1647.21 cm^{-1} absorbance band was contributed to carboxylate group, that specified as a bending mode for absorption of water in fiber. This peak also shows high intensity due to the influence of the interaction between water and cellulose, as well as intermolecular bonds. The 1250–1050 cm^{-1} range peak indicated a strong functional C-O-C group for pyranose and glucose ring skeletal. Similar results were identified by Nguyen et al. [26] shown an increase in cellulose content. The lack of peaks at 1244.09 and 1242.16 cm^{-1} in mercerized cellulose reveal that pectin was removed from the sample. The peak presented in Fig. 4 shows that the compositions of bamboo such as lignin, hemicellulose and pectin affected the properties of polymer nanocomposite film mostly have been eliminated. The chemical treatment efficacy in the sample resulted in the removal of lignin, hemicellulose and pectin. The carbohydrate-lignin linkages breakdown achieved reductions during acid hydrolysis, whereas the hemicelluloses hydrolysis, the formation of cleavage bonds and solubilization of others non-cellulosic components (Table 3).

Table 3 IR Spectra table for fiber components [29]

Fiber component	Wavelength (cm^{-1})	Functional group	Compounds
Cellulose	4000–2995	OH	Acid, methanol
	2890	H-C-H	Alkyl, aliphatic
	1640	O-H	Absorbed water
	1270–1232	C-O-C	Aryl-alkyl ether
	1170–1082	C-O-C	Pyranose ring skeletal
Lignin	4000–2995	O-H	Acid, methanol
	2890	$C\text{-}H_2$	Alkyl, aliphatic
	1730–1700		Aromatic
	1632	$C = C$	Aromatic skeletal mode
	1108	O-H	C-OH
	1613, 1450	$C = C$	Aromatic skeletal mode
	1430	$O\text{-}CH_3$	Methoxyl-$O\text{-}CH_3$
	1270–1232	C-O-C	Aryl-alkyl ether
	1215	C-O	Phenol
	700–900	C-H	Aromatic hydrogen
Hemicellulose	4000–2995	O-H	Acid, methanol
	2890	$C\text{-}H_2$	Alkyl, aliphatic
	1765–1715	$C = O$	Ketone and carbonyl
Pectin	4000–2995	O-H	Acid, methanol
	2890	$C\text{-}H_2$	Alkyl, aliphatic
	1765–1715	$C = O$	Ketone and carbonyl
	1680–1640	$C = C$	Alkene
	1300–1150	C-H	Alkyl halide
	1270–1232	C-O-C	Aryl-alkyl ether

3.2 *Fourier Transform Infrared Spectrophotometer (FTIR) of MCC/MMT/PLA Nanocomposite Film*

The absorption band range from 4000–2995 cm^{-1} was presented the hydroxyl groups stretching variation. This peak was also observed in MCC/MMT/PLA polymer nanocomposite sample at peak strength of 3739.97 cm^{-1} in Fig. 6. The 3739.97 cm^{-1} absorption peak in Fig. 6 indicated a good interaction combination between hydroxyl groups in the cellulose matrix with the hydroxyl and carbonyl groups in the PLA matrix. The 2166.06 cm^{-1} peak intensity in Fig. 4.5 indicated the absorption peak for weak $C = C$ stretching band and 1745.58 and 1082.14 cm^{-1} intensity peak was attributed to the bond of $C = O$ and C-O stretching variations. The 1539.27 and 1031.92 cm^{-1} peak was the variation $-CH_3$ and $-OH$ bending. Arjmandi et al. [30] supported the same findings.

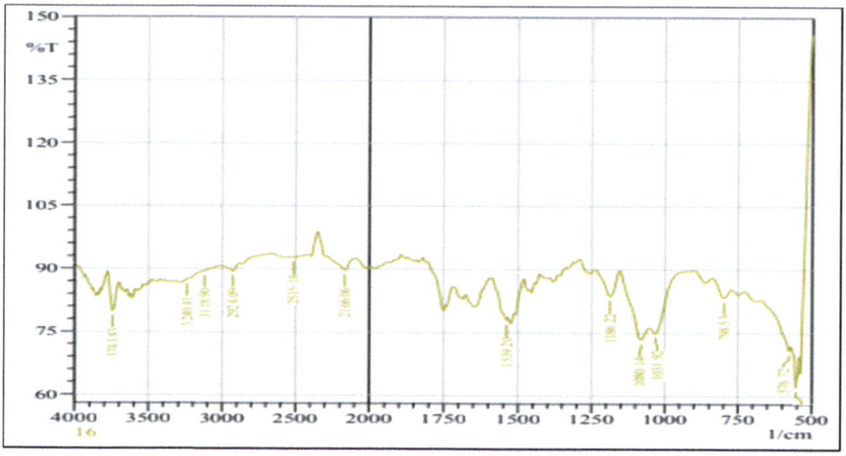

Fig. 6 IR spectra of sample 10 MCC/MMT/PLA polymer nanocomposites

By comparing Figs. 5 and 6, the peaks intensity shifting toward the lower rate from 1757.36,1031.92 and 690.52 cm^{-1} in the PLA sample to 1745.58, 1041 and 686 cm^{-1} in MCC/MMT/PLA polymer nanocomposite, respectively. The shifted absorption peaks observed in Fig. 5 and Fig. 6 were caused by the -OH hydrogen bond formation within the group in the PLA matrix with Si-O bond in MMT [30]. The 576.72 cm^{-1} peak intensity in Fig. 6 shows the Si-O stretching variation bond of the MMT nanofiller. This peak shows a good adhesion interaction between polymer matrix and nanofiller used. This interaction indicated that PLA and MMT particles to cling each other during the preparation of MMC/MMT/PLA polymer nanocomposite film by polar interaction development of functional groups. Arjmandi et al. [30] and Li et al. [31] reported that MMT has several uniform polar sites on its structure that indicate the content of electron density along with the interlayer spaces and their surfaces. The addition of MMT and cellulose (MCC) to the polymer matrix (PLA) does not influence much on the chemical composition, but it decreases the absorption peak of the pure PLA sample, as shown in Fig. 5. This proved by Arjmandi et al. [30] research, whereas the chemical structure of PLA/MMT with cellulose does not change and no new functional group observed. This also shows that the addition of filler does not react chemically, but there was a physical interaction between cellulose and PLA. This observation also reported by Arjmandi et al. [32] and Qu et al. [33]. The increasing of the absorption peak at 3743.83 cm^{-1} in Fig. 6 from peak intensity of 3741.9 cm^{-1} in Fig. 6 was due to the interactions between the –OH groups in cellulose matrix and C-O groups in PLA matrix.

3.3 Bamboo Cellulose Morphological Analysis

The SEM result as in Figs. 7, 8, and 9 indicated the effect of ball milling on the bamboo cellulose. Ball milling was a method used to reduce the size of cellulose from micro-scale powders to nano-scale powders by mechanical shearing. Microfibrils cellulose powder that was treated underwent a ball milling process in a planetary ball mill (PM 100) at a constant milling speed of 200 rpm for 5–7 h in dry conditions. Few factors were considered as these will have effects on the characteristics of the cellulose, include the morphological, structural, thermal stability and crystallinity, i.e. the mass of the milling medium, milling time, and speed of the milling machine.

Fig. 7 The morphology of the bamboo cellulose sample (**a**) non-ball milled, and (**b**) 1st cycle ball milling

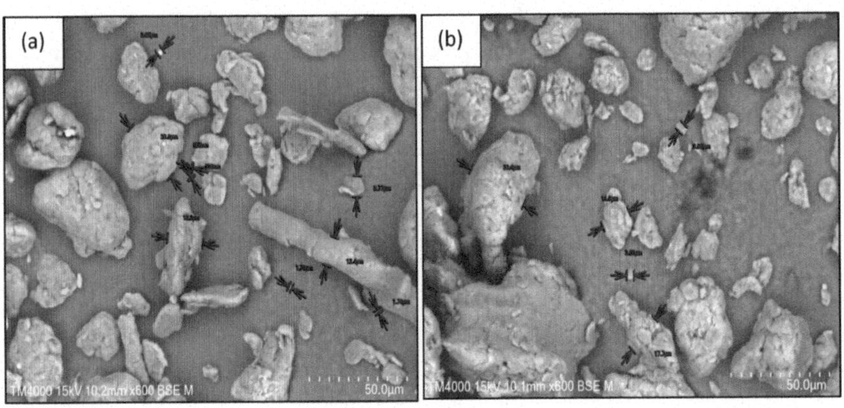

Fig. 8 The morphology of the bamboo cellulose sample (**a**) 2nd cycle milling and (**b**) 3rd cycle ball milling

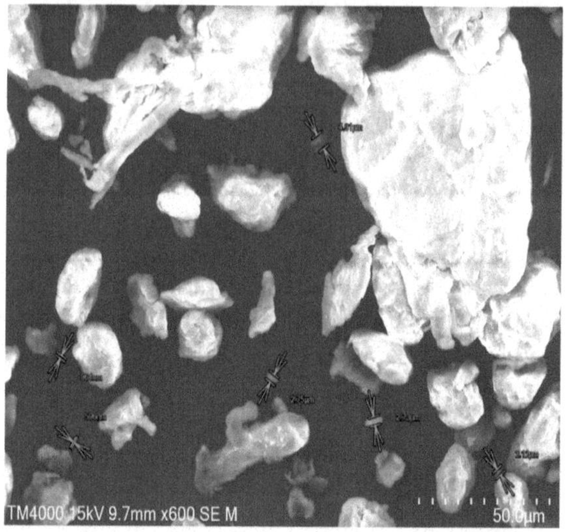

Fig. 9 The morphology of the bamboo cellulose sample for 4th cycle ball milling

Cellulose chains were joined due to interaction of Van der Waals and hydrogen bond, which form fibrils that are assembled into large microfibrils [34]. Figure 7 (a) shows multiple irregular rod-like microfibrils structure before underwent milling process. The average non-ball milled microfibrils diameters were approximately less than 18 μm and mixed with few large diameter fibers. After 7 h of milling process, the microfibrils cellulose rigid compact structure exhibit an intuitive reduction into loosened open-pore width particles roughly 24.7–11.9 μm, as shown in Fig. 7 (b). During the first cycle, cellulose microfibrils seem to be very sensitive to mechanical structure, whereas the irregular rod-like structure changed to flat spherical shape with coarse surfaces. For each milling cycle means prolonged milling time, resulted gradually decrease in particles sizes. This shows that the cellulose powders pulverization occurred during the milling process [24]. However, some particles had a large diameter than non-ball milled microfibrils cellulose. Shashanka and Chaira [35] explained the phenomenon of the soft particle structure that undergoing milling welded together to form a single sizeable lamellar structure.

The energy impact was released from the rotating grinding jar around its axis, initiates attrition between the milling medium stainless-steel balls. The frictional forces caused difference in the rotation speed of stainless-steel balls, whereas the cellulose in the milling vessel release high dynamic energies that effectively increasing the degree of reduction in size. Piras et al. [36] mentioned the milling process efficiency that depends on the cylindrical vessels size, which enables higher kinetic energy and more substantial pressure effects by increasing the distance between cellulose particles. In other words, the size medium level selection of the milling jar also affected the grinding result. It was necessary to provide enough space for the steel balls to collide with enough energy impact to reducing the size.

The second ball milling cycle was carried out for another 7 h at the same process speed to break the large particles size, which was observed in Fig. 8 (a). The particles aspect ratio decreased to produce fragments with a diameter of less than 8 μm and some nanoscale cellulose. The nanocellulose size was within a range of 682–596 nm. Irregular cellulose fragments were the minority longitudinal fragments. After several cycles, it was observed that the effect of the ball milling process on large particles is weakened by the small cellulose particle's accumulation, which prevents reducing in size of cellulose fiber. The 3rd cycle was displayed in Fig. 8 (b). The cellulose particles majority were crushed into small flaky fragments with less than 4 μm size with some predominant nanoparticles sample. Many smaller pores and compact structures were observed on the pulverized particles after the 3rd cycle, due to the longer collision with the milling balls. However, there are still few large sample fragments remained. This was due to the cellulose fiber's inhomogeneity structure. Zheng et al. [24], and Gao et al. [37] observed the microcrystalline cellulose breakage and agglomeration during the mechanical grinding process, whereas the individual cellulose particles began to aggregate. Over milling initiated compact cellulose globular fragments exhibiting granular texture [38].

Figure 9 show some cellulose particles become agglomerated fibers and produced a few large-scale particles, which was caused by the forces of Van der Waals between the particles. Despite the formation of particles due to agglomeration, a few nanoscale particles were observed. This shows that the aggregated particle increases as the hydrogen bond break and begins to form a cleavage on the particle structure after the 4th cycle. As Barbash et al. [39] and Nemazifard et al. [40], indicated that the cellulose particle sizes affected the mechanical properties, as small-size cellulose particles provide high strength for polymer nanocomposite film.

3.4 *Scanning Electron Microscopy (SEM) of MCC/MMT/PLA Nanocomposite Film*

The images illustrated in Figs. 10, 12, and 14 represent the SEM analysis with a magnification of 100, while Figs. 11, 13 and 15 display the SEM images with a magnification of 500 for pure PLA, sample 10 MCC/MMT/PLA and sample 13 MCC/MMT/PLA respectively. Figure 10 shows the fracture surface of a pure PLA that had a smooth surface compared to the nanocomposite surfaces of MCC/MMT/PLA in Figs. 12 and 14. Arjmandi et al. [30] reported that the PLA had a featureless and brittle surface. There were some small white flakes observed in Fig. 10, which originated from the nonmelted PLA powder during the compression process. Figure 10 shows the plastic breakage mechanism, which allows the PLA fibrils formation. As a result of plastic breakage, fibrils split, buckled and fell onto the rupture surface, as shown in Fig. 11.

The cellulose different concentration distribution and MMT in the polymer matrix was observed in Figs. 12 and 14 for the Sample 10 and Sample 13 of MMC/MMT/PLA. The MMC/MMT/PLA cross-sectional surface in Fig. 12 shows an even cellulose dispersion of MMT nanofiller to the PLA. The arrow pointed towards

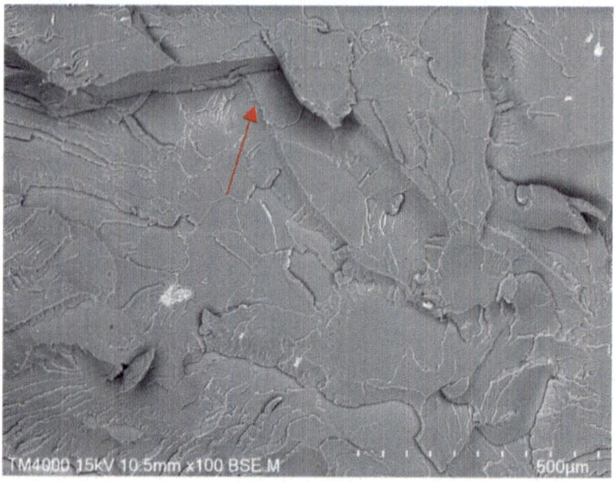

Fig. 10 The morphological image of pure PLA (100 magnification)

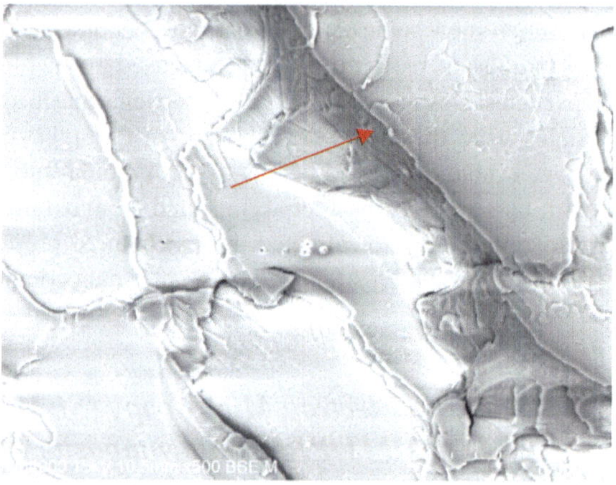

Fig. 11 The morphological image of pure PLA (500 magnification)

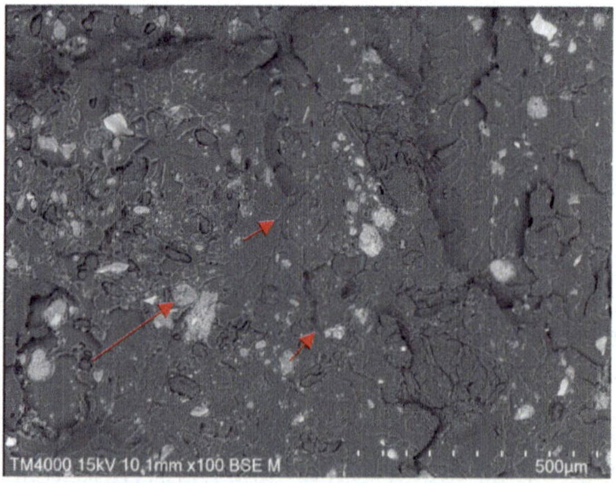

Fig. 12 The morphological image of sample 10 MCC/MMT/PLA (100 magnification)

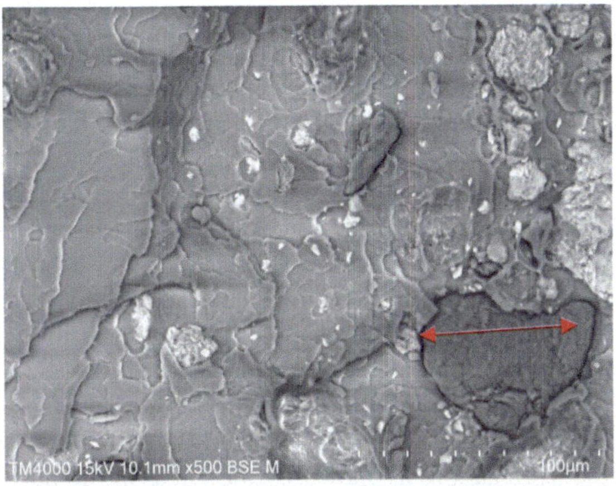

Fig. 13 The morphological image of sample 10 MCC/MMT/PLA (500 magnification)

Fig. 14 The morphological image of sample 13 MCC/MMT/PLA (100 magnification)

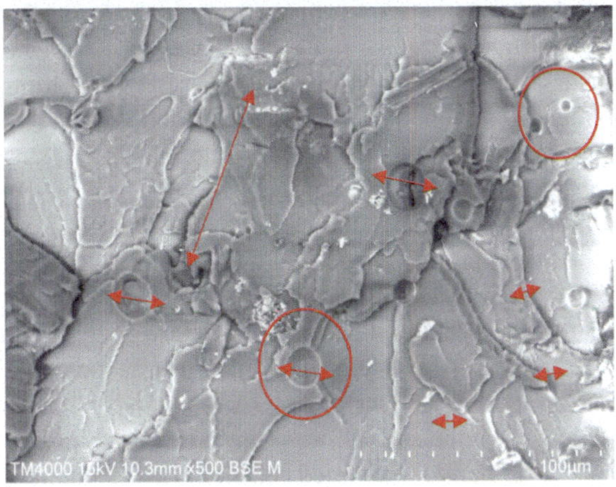

Fig. 15 The morphological image of sample 13 MCC/MMT/PLA (500 magnification)

large particulate in Fig. 12 indicated the cellulose, meanwhile the small one shows MMT nanofiller fragments represent. Arjmandi et al. [30] mentioned that the small polymer nanocomposite fragments in the SEM images are the nanofiller. Figure 12 contain less cellulose and MMT concentration around 4 wt and 2 wt%, respectively, while the sample in Fig. 14 comprised of 5.716 wt% of cellulose and 6.244 wt% of MMT. There was no cellulose aggregation observed in Figs. 12 and 13. Mubarak [41] proved the concentration of cellulose particles was distributed in the matrix and lower than 5 wt%. However, there was one air bubble sac presence in the surface of Sample 10 which was due to the tensile strength reduction of the film production.

Figure 14 shows the cellulose and MMT filler irregular distribution in PLA matrix due to the high amount of cellulose and MMT concentration in the sample. The cellulose distribution was within the circled area in Fig. 14. Dos Santos et al. [42] and Mubarak [41] also explained that cellulose poor distribution caused cellulose aggregation, as shown by the circle in the polymer nanocomposite structure in Fig. 14, which was affecting the tensile properties of the composite. The cellulose poor particulates dispersion in the polymer matrix was due to the hydrophilic cellulose behavior in nature and PLA hydrophobic behavior. Therefore, there was no compatibility between MCC and PLA matrix due to the weak interaction [43]. On the other hand, in Figs. 14 and 15, there are many air bubbles present on the polymer nanocomposite surface, which caused the rough sample surface. The air bubble presence entrapped in the sample influenced the film elasticity [44].

3.5 Energy Dispersive X-Ray (EDX/EDS) Spectroscopy Characterization

The energy dispersive x-ray (EDX/EDS) diffraction was correlated with the SEM result as discussed in the previous section. Figures 16, 17, 18 and 19 show the bamboo cellulose element peaks composition present during the ball milling process, while the EDX diffraction spectra in Fig. 20 shows the MCC/MMT/PLA polymer nanocomposite element composition. Based on EDX spectrum, the higher prominence atom percentages elements were carbon (C) and oxygen (O). This had shown that C and O were the major components of cellulose [45]. The small impurities amount, particularly aluminum (Al) element was also detected by the EDX spectra. By referring to Tables 4, 5, 6, 7 and 8, the C percentage was 59.87, 60.75, 63.58, 61.73 and 60.82% and O percentage were 39.88, 39.02, 35.62, 37.22 and 39.02%, respectively.

By referring to the Al atom percentage, Table 4 holds 0.25% of the total percentage, while Table 5 shows 0.23%, followed by 0.79% in Table 6, and 1.04% in Table 7. There was no Al presence in the polymer nanocomposite in Table 8. The Al spectra was due to the presence of Al_2O_3 compounds in the bamboo structure [46]. In cases of toxicity in the soil, Al was considered too high, as the root elongation inhibited. Zhao and Shen [47] reported cell elongation reduction and cell division due to the Al interference with the cell wall. This occurred as the plant mechanism reduce the Al

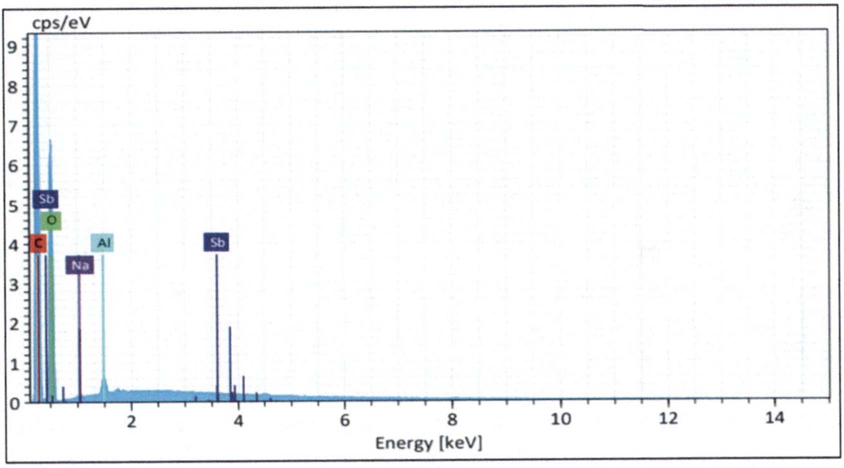

Fig. 16 Energy Dispersive X-ray (EDX) diffraction spectra with its element compositions of bamboo cellulose for 0th cycle

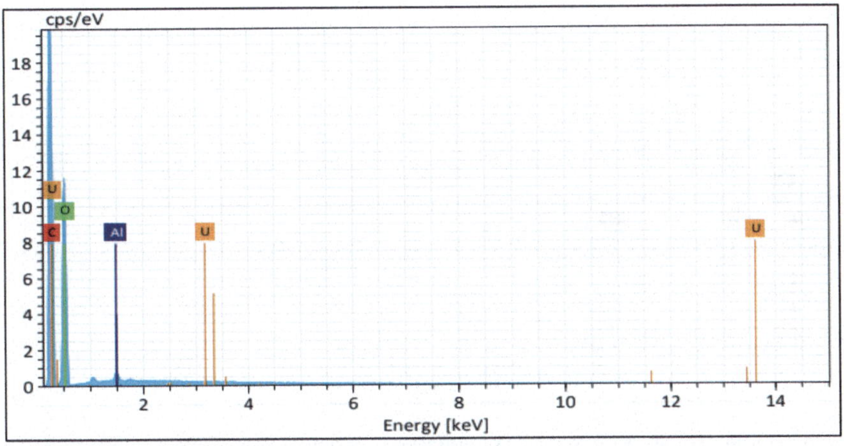

Fig. 17 Energy Dispersive X-ray (EDX) diffraction spectra with its element compositions of bamboo cellulose for 1st cycle

toxification. Since it was the root responsible for the nutrient's uptake from the soil, it tends to halt its growth and reduce the Al toxification into the plant [47]. Based on the observation in Table 8 of MCC/MMT/PLA, there were two small-scale new elements, which are the Nickel (Ni) and Ruthenium (Ru) presence in the polymer nanocomposite. The Ni percentage presence in the composite was 0.14% and 0.02% for Ru (Fig. 18).

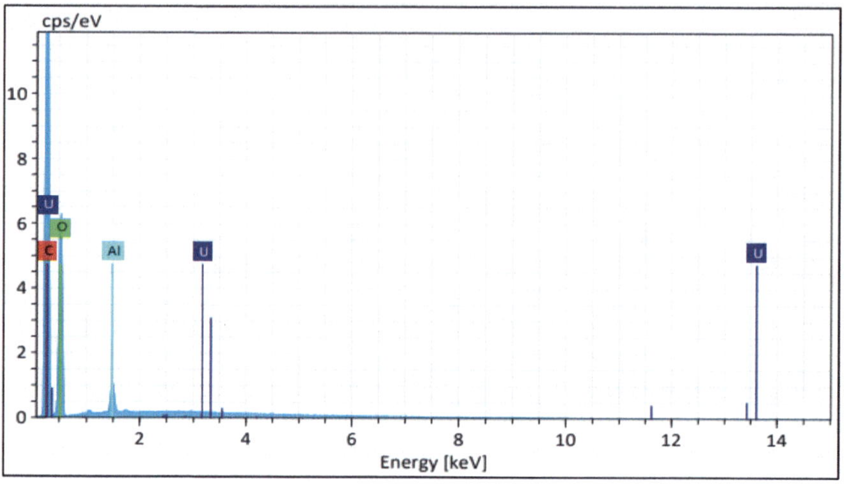

Fig. 18 Energy Dispersive X-ray (EDX) diffraction spectra with its element compositions of bamboo cellulose for 2nd cycle

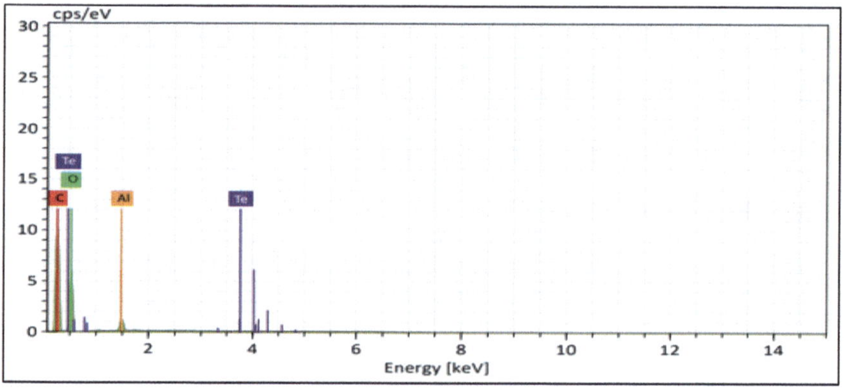

Fig. 19 Energy Dispersive X-ray (EDX) diffraction spectra with its element compositions of bamboo cellulose for 3rd cycle

3.6 Brunauer-Emmett-Teller (BET)

Based on Table 9, the specific surface area was increased in value from the 3rd cycle to the 4th cycle. The cellulose 4th cycle specific surface area was at highest value at 39.708 m^2/g compared to the cellulose 3rd cycle at 33.548 m^2/g. These results were corresponded to the cellulose particles size reduction after ball milling that was shown in Figs. 6 (b) and 7, due to the open particles surface and less particulates merging during the vacuum drying process. The increase of specific surface area caused by lack of coalescence particles. In the polymer nanocomposite with MMT,

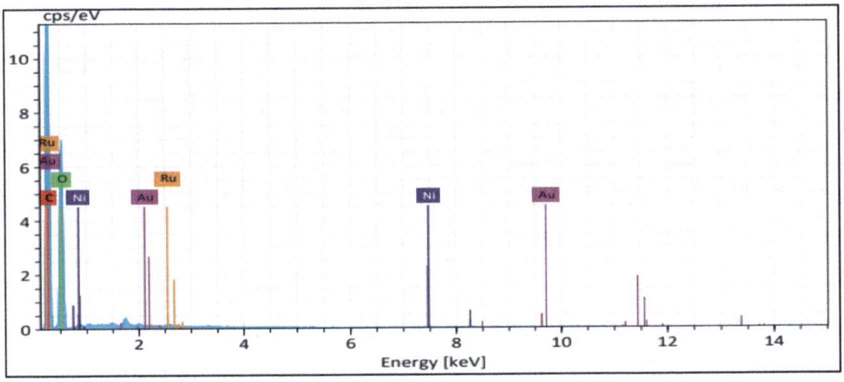

Fig. 20 Energy Dispersive X-ray (EDX) diffraction spectra with its element compositions of MMC/MMT/PLA polymer nanocomposite

Table 4 EDX composition of bamboo cellulose for 0th cycle

Element	Mass (%)	Atom (%)	Abs. error (%)(σ)	Rel. error (%)(σ)
C	52.72	59.87	6.13	11.62
O	46.77	39.88	5.63	12.05
Al	0.50	0.25	0.05	10.47
Sb	0.00	0.00	0.00	27.79
Na	0.00	0.00	0.00	4.18
Sum	100.00	100.00		

Table 5 EDX composition of bamboo cellulose for 1st cycle

Element	Mass (%)	Atom (%)	Abs. error (%)(σ)	Rel. error (%)(σ)
C	53.62	60.75	6.12	11.42
O	45.88	39.02	5.44	11.85
Al	0.45	0.23	0.05	10.97
U	0.05	0.00	0.01	10.32
Sum	100.00	100.00		

Table 6 EDX composition of bamboo cellulose for 2nd cycle

Element	Mass (%)	Atom (%)	Abs. error (%)(σ)	Rel. error (%)(σ)
C	56.20	63.58	6.43	11.45
O	41.94	35.62	5.06	12.06
Al	1.56	0.79	0.10	6.53
U	0.30	0.02	0.04	13.52
Sum	100.00	100.00		

Table 7 EDX composition of bamboo cellulose for 3rd cycle

Element	Mass (%)	Atom (%)	Abs. error (%)(σ)	Rel. error (%)(σ)
C	54.27	61.73	6.26	11.54
O	43.59	37.22	5.24	12.01
Al	2.06	1.04	0.12	6.04
Te	0.08	0.01	0.03	38.31
Sum	100.00	100.00		

Table 8 EDX composition of MMC/MMT/PLA polymer nanocomposite

Element	Mass (%)	Atom (%)	Abs. error (%)(σ)	Rel. error (%)(σ)
C	53.50	60.82	6.63	12.40
O	45.73	39.02	6.01	13.14
Ni	0.60	0.14	0.08	12.71
Ru	0.17	0.02	0.04	23.08
Au	0.00	0.00	0.00	1.73
Sum	100.00	100.00		

Table 9 The specific surface area of 3rd cycle and 4th cycle

Sample	Specific surface area (m^2/g)
3rd cycle	33.548
4th cycle	39.708

the size of cellulose particles also influenced the number of cellulose particulates. Based on research conducted by Mondal [48], the decreased cellulose particle size had affected the particles dispersion in the polymer matrix, which correlates to the increasing amount of specific surface area. The polymer nanocomposite mechanical and tensile properties increased as the specific particles surface area increased as incorporated with MMT [49, 50]. The cellulose particles surface area was correlated with the interfacial area between the cellulose fragments and polymer matrix.

3.7 *Tensile Strength and Tensile Modulus*

The MCC/MMT/PLA polymer nanocomposite mechanical properties were investigated based on the result obtained from the tensile testing. Figures 21 and 22 show the MCC/MMT/PLA nanocomposite tensile strength and Young's modulus. Based on Fig. 21, the addition of cellulose and MMT increased the polymer strength. This was observed by comparing Sample 0 with other samples of polymer nanocomposite. However, there was incorrect correlation between the amount of cellulose and MMT in the composite, whereas there was a reduce in strength of the PLA as shown in

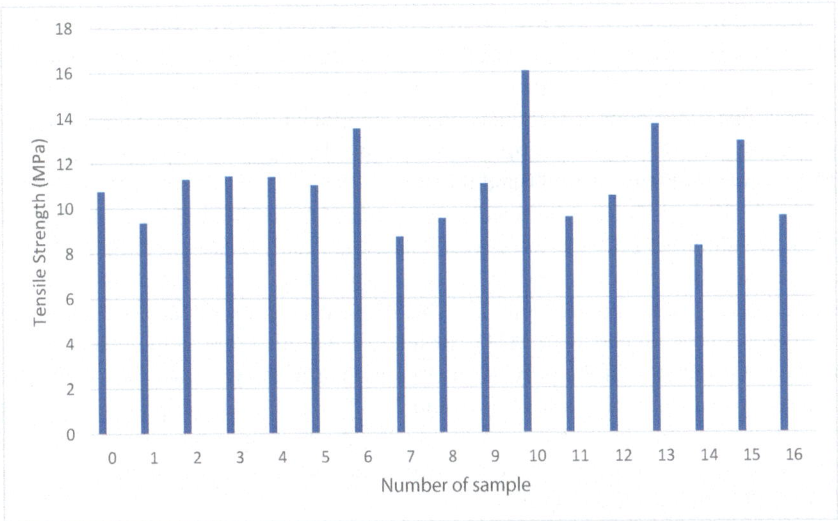

Fig. 21 Tensile strength of pure PLA (sample 0) and MCC/MMT/PLA nanocomposite

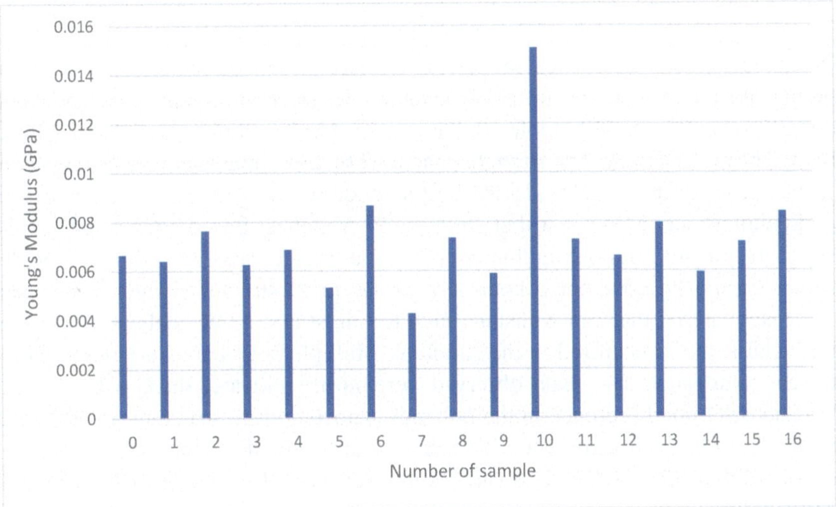

Fig. 22 Young's modulus of pure PLA (sample 0) and MCC/MMT/PLA nanocomposite

Sample 1, 7, 8, 11 and 14. Mubarak [41] also stated the nanofiller addition perfectly improve the composite elongation break. Excessive addition of cellulose and MMT led to particles agglomeration in samples, which initiated brittle behavior, because of the non-uniform stress transfer and stress concentration point in the sample. In addition, the PLA chain mobility was restricted with the cellulose and MMT raising amount that reduce the elongation at break and Young's modulus of the sample. This was proved by Sample 7 in Figs. 21 and 22, which used a high amount of cellulose (6.41 wt%) and low amount of MMT (1 wt%) of fabrication of polymer nanocomposite that resulted in 8.72 MPa of strength and 0.004266GPa of Young's modulus. This was due to the poor MMT and cellulose components dispersion whereas development the particles agglomeration in the sample. The agglomeration prevented the full engagement between MMT and the PLA.

Sample 10 had the highest tensile strength and Young's modulus, which used 4wt% of cellulose, 2.434 wt% of MMT nanofiller and 93.566 wt% of PLA that had a tensile strength of 16.063 MPa, as shown in Fig. 21, and 0.015083 GPa of Young's modulus, as shown in Fig. 22. This PLA polymer excellence stress transfer to cellulose incorporated with MMT. Besides, it was also due to the component's excellence dispersion in the polymer matrix and particles synergic effect [42]. The cellulose hydrogen bond compatibility and the PLA hydroxyl group provided good interaction between the matrixes. Additionally, the high specific surface area was influenced by the surface interaction between the matrixes. The cellulose size also affected the film tensile strength [42].

4 Conclusions

The nanocomposite mechanical property comprised the tensile and modulus strength was depended on the compositions of the materials. The highest tensile and modulus strength achieved was of Sample 10 that contain 4wt% cellulose, 2.434wt% MMT and 93.566wt% of PLA. The strength generated by this composite was 16.063 MPa for the tensile strength and 0.015083 for the modulus strength. This was due to the good cellulose and MMT distribution in the PLA matrix. Based on the SEM analysis, there was no cellulose agglomeration at the fracture area. In addition, Sample 10 nanocomposite does not contain any air sac in its structure, which had made it a dense material that can withstand high tensile stress. FTIR analysis shows the interfacial reaction occurred in the structure. Multiple peak bands are observed in the nanocomposites. The peaks observed were mostly indicated the C = C, C = O, C-O and Si-O bond presence in the polymer nanocomposite structure. In addition, the bamboo fiber extraction was also achieved at for the dewaxing and mercerization of bamboo fiber, which had successfully removed the content of wax, lignin, hemicellulose and pectin.

Acknowledgements The authors would like to acknowledge Universiti Malaysia Sarawak (UNIMAS) for the support.

References

1. Arts, K., Wal, R.v.d., Adams, W.M.: Digital technology and the conservation of nature. Ambio **44**(1), 661–673 (2015). https://doi.org/10.1007/s13280-015-0705-1
2. Owusu, P.A., Asumadu-Sarkodie, S.: A review of renewable energy sources, sustainability issues and climate change mitigation. Cogent Eng. **3**(1), 1167990 (2016). https://doi.org/10.1080/23311916.2016.1167990
3. Siti Suhaily, S., Abdul Khalil, H.P.S., Wan Nadirah, W.O., Jawai, M.: Bamboo based biocomposites material, design and applications. In: Mastai, Y. (ed.) Materials science: advanced topics, intechopen, pp. 1–10 (2013). https://doi.org/10.5772/56057
4. Liu, Y., Ying, Z., Wang, S., Liao, J., Lu, H., Ma, L., Li, Z.: Modeling the impact of reproductive mode on masting. Ecol. Evol. **7**(16), 6284–6291 (2017). https://doi.org/10.1002/ece3.3214
5. Awoyera, P.O., Ugwu, E.I.: Sustainability and recycling of bamboo for engineering applications. Encycl. Renew. Sustain. Mater. **2**(1), 337–346 (2017). https://doi.org/10.1016/B978-0-12-803581-8.10367-4
6. Ng, F., & Noor, A. (1980). Malaysia. In: Lessard, G., Chouinard, A. (eds.) Bamboo research in Asia. Proceedings of the workshop. Singapore: International Union Forestry Research Organization and International Development Research Centre, Singapore, pp. 91–96. http://hdl.handle.net/10625/5559
7. Kamaruzaman, A.B.: Bamboo resource in Peninsular Malaysia. Bul. Buluh (Bamboo) **1**(1), 8–9 (1992)
8. Wong, K.M.: Current and potential use of bamboo in Peninsular Malaysia. J. Am. Bamboo Soc. **7**(1&2), 1–15 (1989)
9. Mohamed, A. H., & Appanah, S.: Bamboo resources conservation and utilisation in Malaysia. In: Rao, A.N., Rao, V.R. (eds.) Bamboo—conservation, diversity, ecogeography, germplasm, resource utilization and taxonomy (1998). https://www.bioversityinternational.org/fileadmin/bioversity/publications/Web_version/572/ch27.htm
10. Roslan, S.A.H., Rasid, Z.A., Hassan, M.Z.: Bamboo reinforced polymer composite—a comprehensive review. IOP Conf. Ser.: Mater. Sci. Eng. **344**(1), 012008 (2018). https://doi.org/10.1088/1757-899X/344/1/012008
11. Liew, F.K., Hamdan, S., Rahman, M.R., Rusop, M.: Thermomechanical properties of jute/bamboo cellulose composite and its hybrid composites: the effects of treatment and fiber loading. Adv. Mater. Sci. Eng. **2017**(1), 1–15 (2017). https://doi.org/10.1155/2017/8630749
12. Sun, S., Sun, S., Cao, X., Sun, R.: The role of pretreatment in improving the enzymatic hydrolysis of lignocellulosic materials. Biores. Technol. **199**(1), 49–58 (2016). https://doi.org/10.1016/j.biortech.2015.08.061
13. Menon, M.P., Selvakumar, R., Kumar, P.S., Ramakrishna, S.: Extraction and modification of cellulose nanofibers derived from biomass for environmental application. RSC Advances **7**(68), 42750–42773 (2017). https://doi.org/10.1039/C7RA06713E
14. Kang, H., Liu, R., Huang, Y.: Cellulose-based gels. Macromol. Chem. Phys. **217**(2), 1322–1334 (2016). https://doi.org/10.1002/macp.201500493
15. Mattausch, H. (2015). Properties and applications of nanoclay composites. In: Laske, S. (ed.) Polymer nanoclay composites, William Andrew, 127–155. https://doi.org/10.1016/B978-0-323-29962-6.00005-4
16. Shetti, N., Nayak, D., Reddy, K., Aminabhvi, T.: Graphene-clay-based hybrid nanostructures for electrochemical sensors and biosensors. In: Pandikumar, A., Rameshkumar, P. (eds.) Graphene-Based Electrochem. Sens.S Biomol. **1**(1), 235–274 (2019). https://doi.org/10.1016/B978-0-12-815394-9.00010-8
17. Ghadiri, M., Chrzanowski, W., Rohanizadeh, R.: Biomedical applications of cationic clay minerals. RSC Advances **5**(37), 29467–29481 (2015). https://doi.org/10.1039/C4RA16945J
18. Rahman, M.R., Hui, J.L., Hamdan, S.: Introduction and reinforcing potential of silica and various clay dispersed nanocomposites. In: Rahman, M.R. (ed.) Silica and clay dispersed polymer nanocomposite. Woodhead Publishing, Cambridge, pp. 1–24 (2018). https://doi.org/10.1016/B978-0-08-102129-3.00001-4

19. Uddin, F.: Montmorillonite: An introduction to properties and utilization. In: Current topics in the utilization of clay in industrial and medical applications, IntechOpen (2018). https://doi.org/10.5772/intechopen.77987
20. ASTM E1252-98.: Standard practice for general techniques for obtaining infrared spectra for qualitative analysis, ASTM International, West Conshohocken, PA (2013). https://doi.org/10.1520/E1252-98R13E01
21. ASTM E168-16.: Standard practices for general techniques of infrared quantitative analysis, ASTM International, West Conshohocken, PA (2016). https://doi.org/10.1520/E0168-16
22. ASTM E2015-04.: Standard guide for preparation of plastics and polymeric specimens for microstructural examination, ASTM International, West Conshohocken, PA (2014). https://doi.org/10.1520/E2015-04R14
23. ASTM D638-14.: Standard test method for tensile properties of plastics, ASTM International, West Conshohocken, PA. https://doi.org/10.1520/D0638-14
24. Zheng, Y., Fu, Z., Li, D., Wu, M.: Effects of ball milling processes on the microstructure and rheological properties of microcrystalline cellulose as a sustainable polymer additive. Material **11**(7), 1057 (2018). https://doi.org/10.3390/ma11071057
25. Radotic, K., Micic, M.: Methods for extraction and purification of lignin and cellulose from plant tissues. In: Micic, M. (ed.) Sample preparation techniques for soil, plant, and animal samples, pp. 365–376 (2016). https://doi.org/10.1007/978-1-4939-3185-9_26
26. Nguyen, H.D., Mai, T.T., Nguyen, N.B., Dang, T.D., Le, M.L., Dang, T.T., Tran, V.M.: A novel method for preparing microfibrillated cellulose from bamboo fibers. Adv. Nat. Sci.: Nanosci. Nanotechnol. **4**(9), 015016 (2013). https://doi.org/10.1088/2043-6262/4/1/015016
27. Gabhane, J., William, S., Vaidya, A., Das, S., Wate, S.: Solar assisted alkali pretreatment of garden biomass: effects on lignocellulose degradation, enzymatic hydrolysis, crystallinity and ultra-structural changes in lignocellulose. Waste Manag. **40**(1), 92–99 (2015). https://doi.org/10.1016/j.wasman.2015.03.002
28. Li, D., Long, L., Ding, S.: Alkaline organocolv pretreatment of different sorghum stem parts for enhancing the total reducing sugar yield and p-coumaric acid release. Biotechnol. Biofuels **12**(106), 1–10 (2020). https://doi.org/10.1186/s13068-020-01746-4
29. Moran, J.I., Alvarez, V.A., Cyras, V.P., Vasquez, A.: Extraction of cellulose and preparation of nanocellulose from sisal fiber. Cellulose **15**(1), 149–159 (2008). https://doi.org/10.1007/s10570-007-9145-9
30. Arjmandi, R., Hassan, A., Mohamad Haafiz, M., Zakaria, Z.: Effect of micro—And nanocellulose on tensile and morphological properties of montmorillonite nanoclay reinforced polylactic acid nanocomposite. In: Jawaid, M., Qaiss, A., Bouhfid, R. (eds.) Nanoclay reinforced polymer composites, pp. 103–125 (2016). https://doi.org/10.1007/978-981-10-0950-1_5
31. Li, B.L., Setyawati, M.L., Chen, L., Xie, J., Ariga, K., Lim, C.-T., Garaj, S., Leong, D.T.: Directing assembly and disassembly of 2D MoS2 nanosheets with DNA for drug delivery. Appl. Mater. Interfaces **9**(8), 15286–15296 (2017). https://doi.org/10.1021/acsami.7b02529
32. Arjmandi, R., Hassan, A., Haafiz, M.M., Zakaria, Z.: 7—Effects of cellulose nanowhiskers preparation methods on the properties of hybrid montmorillonite/cellulose nanowhiskers reinforced polyactic acid nanocomposites. In: Lau, K.-T., Hung, A.P.-Y. Natural fiber—reinforced biodegradable and bioresorbable polymer composites, Woodhead Publishing, Cambridge, pp. 111–136 (2017). https://doi.org/10.1016/B978-0-08-100656-6.00007-8
33. Qu, P., Gao, Y., Wu, G., Zhang, L.: Nanocomposite of poly (lactic acid) reinforced with cellulose nanofibrils. BioResources **5**(3), 1811–1823 (2010). https://ojs.cnr.ncsu.edu/index.php/BioRes/article/view/BioRes_05_3_1811_Qu_GWZ_Nanocomposites_PLA_Cellulose
34. Hamad, W.Y.: Cellulose nanocrystals: properties, production and applications, pp. 1–312. John Wiley & Sons, West Sussex (2017)
35. Shashanka, R., Chaira, D.: Ball milled nano-structured stainless steel powders: fabrication and characterisation. Educreation Publishing, New Delhi (2017)
36. Piras, C.C., Prieto, S.F., De Borggraeve, W.M.: Ball milling: a green technology for the preparation and functionalisation of nanocellulose derivatives. Nanoscale Adv. **1**(3), 937–947 (2019). https://doi.org/10.1039/C8NA00238J

37. Gao, C., Xiao, W., Ji, G., Zhang, Y., Cao, Y., Han, L.: Regularity and mechanism of wheat straw properties change in ball milling process at cellular scale. Biores. Technol. **241**(1), 214–219 (2017). https://doi.org/10.1016/j.biortech.2017.04.115
38. Vaidya, A.A., Donaldson, L.A., Newman, R.H., Suckling, I.D., Campion, S.H., Lloyd, J.A., Murton, K.D.: Micromorphological changes and mechanism associated with wet ball milling of Pinus radiata substrate and consequences for saccharification at low enzyme loading. Biores. Technol. **214**(1), 132–137 (2016). https://doi.org/10.1016/j.biortech.2016.04.084
39. Barbash, V., Yaschenko, O., Alushkin, S., Kondratyuk, A., Posudievsky, O., Koshechko, V.: The effect of mechanochemical treatment of the celllulose on characteristics of nanocellulose films. Nanoscale Res. Lett. **11**(1), 410 (2016). https://doi.org/10.1186/s11671-016-1632-1
40. Nemazifard, M., Kavoosi, G., Marzban, Z., Ezedi, N.: Physical, mechanical, water binding and antioxidant properties of cellulose dispersions and cellulose film incorporated with pomegranate seed extract. Int. J. Food Prop. **20**(2), 1501–1514 (2017). https://doi.org/10.1080/10942912.2016.1219369
41. Mubarak, Y.: Tensile and impact properties of microcrystalline cellulose nanoclay polypropylene composites. Int. J. Polym. Sci. **2018**(1), 1708695 (2018). https://doi.org/10.1155/2018/1708695
42. Dos Santos, F., Lulianelli, G., Tavares, M.: Effects of microcrystalline and nanocrystals cellulose fillers in materials based on PLA matrix. Polym. Testing **61**(1), 280–288 (2017). https://doi.org/10.1016/j.polymertesting.2017.05.028
43. Shamsabadi, M., Behzad, T., Bagheri, R., Nari-Nasrabadi, B.: Preparation and characterisation of low-density polyethylene/thermoplastic starch composites reinforced by cellulose nanofibers. Polym. Compos. **36**(12), 2309–2316 (2015). https://doi.org/10.1002/pc.23144
44. Hwang, H., Kim, D.-G., Jang, N.-S., Kong, J.-H., Kim, J.-M.: Simple method for high—performance stretchable composite conductors with entrapped air bubbles. Nanoscale Res. Lett. **270**(1), 64–76 (2016). https://doi.org/10.1016/j.apsusc.2012.12.083
45. Kian, L.K., Jawaid, M., Ariffin, H., Alothman, O.Y.: Isolation and characterisation of microcrystalline cellulose from roselle fibers. Int. J. Biol. Macromol. **103**(1), 931–940 (2017). https://doi.org/10.1016/j.ijbiomac.2017.05.135
46. Aminullah, Rohaeti, E., Yuliarto, B., Irzaman.: Reduction of silicon dioxide from bamboo leaves and its analysis using energy dispersive x-ray and fourier transform infrared. IOP Conf. Ser.: Earth Environ. Sci. **209**(1),012048 (2018). https://doi.org/10.1088/1755-1315/209/1/012048
47. Zhao, X.Q., Shen, R.F.: Aluminium-nitrogen interactions in the soil-plant system. Plant Sci. **9**(1), 807 (2018). https://doi.org/10.3389/fpls.2018.00807
48. Mondal, S.: Review on nanocellulose polymer nanocomposites. Polym. Plast. Technol. Eng. **57**(13), 1377–1391 (2017). https://doi.org/10.1080/03602559.2017.1381253
49. Kowalczyk, M., Piorkowska, E., Kulpinski, P., Pracella, M.: Mechanical and thermal properties of PLA composites with cellulose nanofibers and standard size of fibers. Compos. A Appl. Sci. Manuf. **42**(10), 1509–1514 (2011). https://doi.org/10.1016/j.compositesa.2011.07.003
50. Irvin, C.W., Satam, C.C., Meredith, C., Shofner, M.L.: Mechanical reinforcement and thermal properties of PVA tricomponent nanocomposites with chitin nanofibers and cellulose nanocrystals. Compos. A Appl. Sci. Manuf. **116**(1), 147–157 (2019). https://doi.org/10.1016/j.compositesa.2018.10.028

Bamboo Nanocellulose Reinforced Polylactic Acid Nanocomposites

Md Rezaur Rahman and Muhammad Khusairy Bin Bakri

Abstract This chapter discover bamboo nanocellulose nanocomposites. The nanocomposites were characterized by Fourier transform infrared (FTIR) spectroscopy, scanning electron microscopy (SEM) and energy dispersive x-ray (EDS/EDX), respectively. Mechanical test was carried out by universal testing machine. This peak intensity was related to the glycosidic –C1-H deformation, a ring vibration, and –O-H group bending, which attributed to the β-glycosidic linkages between the anhydro glucose units in cellulose. The fractured cross-sectional surface showed that the cellulose fibers only physically interact with the PLA matrix and not chemically interact as confirmed by the chemical structure analysis of FT-IR. The spectrums peaks show that elements of carbon, oxygen and aluminum were detected from the EDX analysis.

Keywords PLA · Bamboo · Nanocomposites · Reinforced · Reinforcement

1 Introduction

Bamboo received a growing attention over the last few decades for its economic and environmental values [1]. Either it is in Asia, Africa, and Latin America, bamboo is usually associated with the indigenous culture and widely known to be used for construction, forestry, agroforestry, and utensils [2]. For a wide range of uses and application Bamboo environmental and physical properties is an exceptional economic resource [3]. According to Bhonde et al. [4], bamboos growth rate was among the fastest-growing plants in the world, whereas it can grow up to 60 cm or more in a day. In nature, not only bamboos belong to the grass family, they are known to have a columnar characteristic rather than tapering [4].

Bamboo can withstand the strong wind and pressure and reached a height up to 40 m [4, 5]. It possesses a few advantages, i.e. lightweight, flexible, tough, higher tensile, and cheap [4]. It also had a huge social, economic and cultural significances

M. R. Rahman (✉) · M. K. B. Bakri
Faculty of Engineering, Universiti Malaysia Sarawak, Jalan Datuk Mohammad Musa, 94300, Kota Samarahan, Sarawak, Malaysia
e-mail: rmrezaur@unimas.my

M. R. Rahman (ed.), *Bamboo Polymer Nanocomposites*, Engineering Materials,
https://doi.org/10.1007/978-3-030-68090-9_8

in East Asia and South East Asia, as it was used extensively for building materials, food sources, and highly versatile raw material and merchandises. In new research supported by Chaowana [6] bamboo had received growing attention in the twentieth century, for its growing nature, high productivity, quick maturity and high strength. The bamboo development was also advance in processing technology and increased in market demand for industrial application, especially as raw material for wood-based composites such as particleboard (PB), medium-density fiberboard (MBF), etc. [6]. Besides, bamboo has been processed into an extended diversity of product ranging from domestic household, musical instrument and weapons [7].

Lobovikov et al. [8] have mentioned the similar opinion where bamboo has been becoming popular as an excellent substitute for wood in production of pulp, paper, board and charcoal. It is often that the bamboo was cultivated outside the forest on farms, where it is more easily managed. The processing normally required a basic skilled labor and started by rural poor communities at a minimal cost. For some reasons, it could offer income-earning opportunities to a significant number or people. Therefore, this chapter investigated the bamboo nanocellulose reinforced polylactic acid nanocomposites. The samples were characterized using Fourier transform infrared spectroscopy (FTIR), scanning electron microscopy (SEM), energy dispersive energy (EDX/EDS) and mechanical test.

2 Methodology

2.1 *Materials*

The raw green bamboo sample was collected from forest at Kota Samarahan, Sarawak. The properties of the materials used were tabulated in Table 1. Polylactic acid, toluene, ethanol, hydrogen peroxide, acetic acid, sodium hydroxide, dichloromethane, montmorillonite (MMT), titanium (IV) oxide, deionized water was used and supply by Sigma Aldrich (US) Inc.

2.2 *Experimental Procedure*

The experimental methodology was divided into four main parts, which consist of preparation of bamboo fiber, preparation of cellulose from bamboo fiber, preparation of cellulose by mechanical fibrillation, and preparation of nanocomposite by using bamboo nanocellulose with polylactic acid matrix.

Table 1 Properties of the materials

Material/Chemical	Chemical formula	Molar mass (g/mol)	State at room temperature	Melting point (°C)	Boiling point (°C)
Polylactic Acid	$(C_3H_4O_2)_n$	60.00	Solid	160.00	-
Toluene	C_7H_8	92.14	Liquid	−95.00	111.00
Ethanol	C_2H_5OH	46.07	Liquid	−114.00	78.00
Hydrogen Peroxide	H2O2	34.01	Liquid	−0.43	150.00
Acetic Acid	CH_3COOH	60.05	Liquid	16.60	118.10
Sodium Hydroxide	NaOH	40.00	Solid	-	1390.00
Dichloromethane	CH_2Cl_2	84.93	Liquid	−96.70	39.60
Titanium (IV) Oxide	TiO_2	79.90	Solid	1843.00	2972.00
Deionized Water	H2O	18.02	Liquid	0.00	100.00
Montmorillonite (MMT)	$(Na,Ca)_{0.33}(Al,Mg)_2(Si_4O_{10})$	-	Solid	-	-

2.3 *Preparation of Bamboo Fiber*

The raw green bamboo underwent grounding and drying process to transform into the fibers form that was used for synthesis of cellulose. 1 m culm of green bamboo was prepared. The green bamboo was grounded by using a mixer grinder to produced chips and powders form, which excluding the internode part. The chips and powder form of bamboo underwent drying process in an oven for 72 h at 70 °C. The dried sample were grounded and sieved by using 600 μm size sieve. The 600 μm mesh size bamboo fibers were used for the synthesis of the cellulose. This sample was labelled as Green Bamboo Fiber (GBF) as shown in Fig. 1.

2.4 *Preparation of Cellulose by Chemical Treatment*

The GBF underwent several step and chemical process to complete the cellulose extraction.

2.4.1 Dewaxing of Bamboo Fiber

A 400 ml of Toluene and 200 ml of ethanol solution was prepared. This solution was put in a round flask to produced toluene ethanol of 2:1 ratio. A round flask filled with water was put on the heating mantel and the temperature was maintained at 100 °C by using the digital thermometer. A Soxhlet extractor was placed on the top of the

Fig. 1 Green Bamboo Fiber (GBF)

boiling round flask and fixed tightly by using the retort stand. 10 g of GBF were weighted and scoop into the cellulose thimble and was placed into the membrane tube of the Soxhlet extractor. A Liebig condenser was placed on top of the extractor as shown in the Fig. 2 the setup of the equipment for dewaxing treatment of bamboo fibers. The process was run until the mixture color disappear which approximately for 1 hour and 6–9 cycles of extraction. The extraction thimble was taken out by using a tweezer after the dewaxing treatment complete. The samples were poured into a beaker and stirred using a glass-rod, while adding the toluene ethanol solution. The mixture was filtered by using the filter tunnel and filter paper, the final product will be distributed evenly on the filter paper by using glass-rod. The final product was left overnight in an oven drying at 70 °C and labelled as Dewaxed Bamboo Fiber (DBF) as in the Fig. 3.

2.4.2 Delignification of Bamboo Fiber

The delignification solutions were prepared by using 82.3 g of Hydrogen peroxide, 106.2 g of Acetic acid in the present of Titanium (IV) Oxide as catalyst. 30 g of dry DBF was weighted and immersed into the delignification solution in a beaker as shown in Fig. 4. The mixture was heated up to 130 °C for 2 h, and the heater was turned off to let it cool at room temperature. The treated product was filtered using Buchner flask and rinsed with deionized water until it reached the pH 7 and undergo

Fig. 2 Dewaxing Step

drying process at 70 °C for 24 h. The dried sample was placed in a bottle and kept in a dark and cool place for next alkaline treatment and labelled as Delignified Bamboo Fiber (DLBF) as in Fig. 5.

2.4.3 Mercerization

The DLBF sample was immersed into the alkaline solution of 6wt% of Sodium Hydroxide to dissolve the pectin and hemicellulose of the fiber at room temperature. An auto shaker equipment as in Fig. 6 was used to stir the mixture at 150 rpm, heated to 80 °C for 2 h and stopped after 8 h of stirring. The mixture was rinsed continuously with deionized water until the pH 7 was reached. The treated product was filtered again by Buchner flask, rinse by using deionized water to reach pH7 and freeze dried at –85 °C for 48 h. The treated product was kept for the next mechanical process and labelled as Bamboo Cellulose Fiber (BCF) as shown in the Fig. 7.

Fig. 3 Dewaxed Bamboo Fiber (DBF)

2.5 Preparation of Cellulose Fibers by Mechanical Process

The mechanical processes were applied to transform the Bamboo Cellulose Fiber (BCF) size into a smaller scale of cellulose fibers. The BCF were grounded by using the mixer grinder for 10 mins. and seven cycle to give the cellulose in powdered form as shown in Fig. 8.

2.6 Preparation of Nanocomposite Film by Using Bamboo Nanocellulose with Polylactic Acid

This process was essential as the cellulose fiber added with the polylactic acid with addition of montmorillonite (MMT) as filler to produce the nanocomposites. In this section, solution casting technique were used to produce nanocomposite film by using dichloromethane (DCM) as the solution. 85wt% of cellulose fiber suspension in 100 ml of DCM was stirred in a beaker. When the cellulose fiber thoroughly dispersed in the solvents, mixture of 1.5wt% of MMT and 6.5wt% of PLA was gradually added the suspension. After the complete dissolution of PLA and MMT,

Fig. 4 Delignification step

the stirring was kept for 2 h. Then, the mixture was poured into the petri dish and the solvents was evaporated under the room temperature for 12 h as shown in Fig. 9. The mixture formulation of nanocomposites was followed by its designed using the Experts Design software for response surface methodology. The mixture formulation was previously tabulated as in the Table 3.

2.7 Experimental Design and Formulation

The design of experiment software Design Expert® 11 were used in designing the formulation of composition and analysis of the experimental data. The parameters used are PLA load, bamboo cellulose load and MMT load at 2-level factorial as shown in the Table 2.

From the software for design of the experiment, 15 runs of experiment were carried as shown in the Table 3 below.

Fig. 5 Delignified Bamboo Fiber (DLBF)

2.8 *Characterization Test*

2.8.1 Fourier Transform Infrared Spectroscopy (FTIR)

The number of scans were set, and the wavelength range taken were between 4000–400 cm^{-1}. The spectrograms were used to analyze the peaks corresponding to the organic compound's chemical groups and compared with the databases to identify the materials from the chemical composition. The test was done according to ASTM E168-16 [9], and ASTM E1252-98 [10] standards.

2.8.2 Scanning Electron Microscopy (SEM)

The morphological images of the broken samples were test using SEM (Hitachi TM3030). The sample was mounted on the holder and metal coated with an ultrathin film of gold with a sputter coater. The images were taken at different magnifications. The tests were conducted according to the ASTM E2015-04 [11] standard.

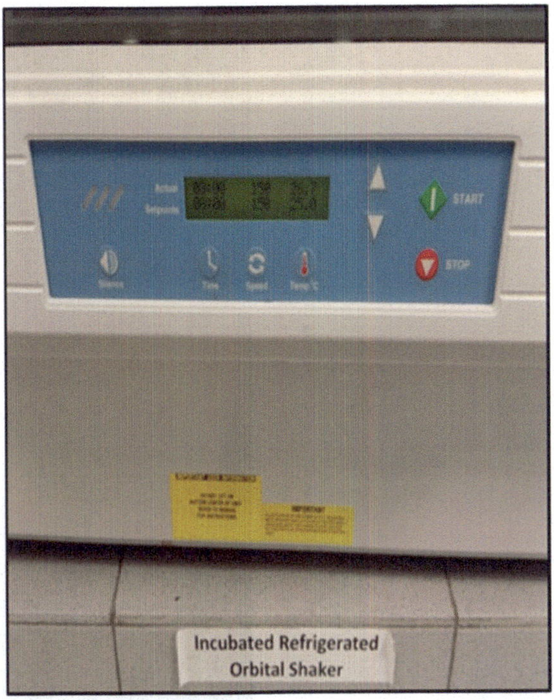

Fig. 6 Mercerization step

2.8.3 Energy Dispersive X-Ray (EDX)

EDX analysis was an x-ray technique used to identify the element composition that present in the sample or specimen. This analysis systems were attachments to the electron microscopy instruments (used for SEM) where the imaging capability of the microscope identifies the specimen elemental composition. The procedure on conducting EDX analysis were similar as SEM analysis as the system is attached in the same instrument. The analysis of EDX was observed from the energy spectrum obtained from the EDX system. The tests were conducted according to the ASTM E1508-12 [12] standards.

2.9 Tensile Test

The tensile properties were obtained, and the test was carried out according to ASTM D638-14 [13] standard on Universal Testing Machine Shimadzu MSC-5/500 at the crosshead speed of 5 mm/min.

Fig. 7 Bamboo Cellulose Fiber (BCF)

Fig. 8 Powder bamboo cellulose fiber

Fig. 9 Nanocomposite film in petri dish

Table 2 Parameters use in the design of experiment

Parameters	Levels	
	Low (−1)	High (+1)
PLA Load	80	90
Bamboo Cellulose Load	5	8
MMT Clay Load	1	2

Table 3 Experimental values from design expert 11

Run	PLA Load (wt%)	Bamboo Cellulose Load (wt%)	MMT Load (wt%)
1	85	6.5	1.5
2	90	5	2
3	80	8	1
4	76.591	6.5	1.5
5	80	5	2
6	85	6.5	2.3409
7	85	6.5	0.659104
8	80	5	1
9	85	9.02269	1.5
10	90	8	1
11	80	8	2
12	93.409	6.5	1.5
13	90	8	2
14	90	5	1
15	85	3.97731	1.5
PLA	100	0.0	0.0

3 Results and Discussions

3.1 Fourier Transform Infrared Spectroscopy Analysis

Figure 10 shows FTIR spectra of raw green bamboo fiber (GBF) and bamboo cellulose fiber (CF) produced through both chemical and mechanical treatments of raw green bamboo.

FTIR spectra revealed the composition changes in the raw green bamboo fiber (GBF) after underwent chemical treatments to produce bamboo cellulose fiber (CF). Figure 10 shows that the red line represents the GBF spectrum, while the black line represents the CF spectrum. The absorption peak of the GBF spectrum at 1508.33 cm^{-1} was attributed to the C = C stretching vibration in the aromatic ring of lignin. While for the CF spectrum, it did not show the C = C stretching vibration in the same region. According to Liew et al. [14], similar finding was found, which indicates the chemical treatment of delignification successfully removed the lignin in the fiber.

The absorption peak intensity at 900.76 cm^{-1} in the CF spectrum was almost to same as the CF spectrum of 900 cm^{-1} obtained from the study conducted by Liew et al. [14]. This peak intensity was related to the glycosidic –C1-H deformation, a ring vibration, and –O-H group bending, which attributed to the β-glycosidic linkages between the anhydro glucose units in cellulose. The rise of intensity peak at 1028.06 cm^{-1} shows that the cellulose content increased by the alkaline treatment. Based on this spectrum's comparison and studies, it was confirmed that cellulose

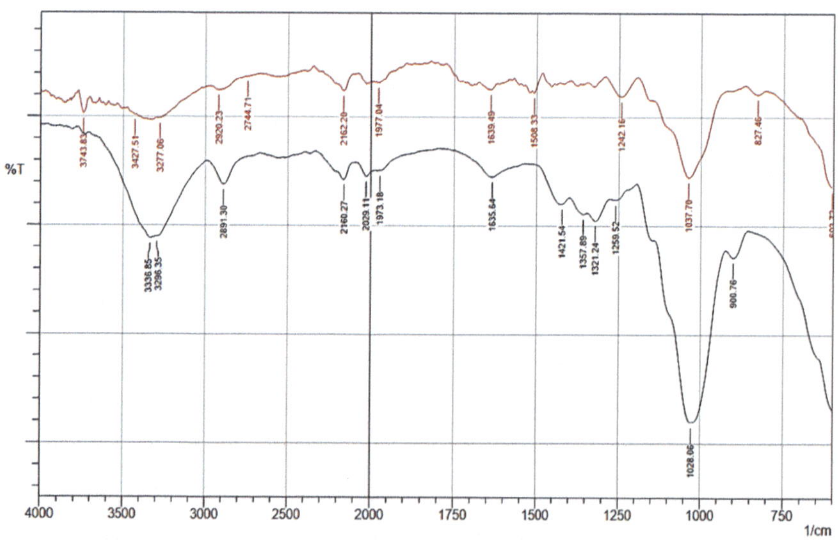

Fig. 10 FT-IR spectrum of (**a**) green bamboo fiber (Red), and (**b**) bamboo cellulose fiber (Black)

fiber was successfully obtained from the raw green bamboo fiber through chemical treatments.

Figure 11 above shows the blue line in the FTIR spectra that illustrated the spectrum for neat PLA, while the green line illustrated the FTIR spectra for nanocomposite film produced through the solution casting technique, which consists of PLA matrix, MMT filler and CF filler with different composition formulation. As shown in the pure PLA spectrum above, the intensity peaks at 1751.38 and 1082.07 cm^{-1} were attributed to the stretching vibration of the C = O functional group and C-O functional group, respectively. This finding was also supported by a similar observation obtained from a study conducted by Arjmandi et al. [15], whereas the absorption peaks for those functional groups were 1764 and 1089 cm^{-1}, respectively. In addition, they also analyze the intensity absorption peaks at 1041.6 and 1452.40 cm^{-1} were attributed to –OH- and –CH3- bending vibrations as obtained in their pure PLA spectrum. This observation was confirmed by the result obtained from the EDS analysis of pure PLA film.

As shown in the nanocomposite film spectrum above, two new peaks had appeared at 480.28 and 542.00 cm^{-1}, which was related to the stretching vibrations of Si–O functional groups contributed by MMT filler. A similar finding also obtained by Arjmandi et al. [15] upon the addition of MMT filler into the PLA matrix. These changes in spectra was due to the good interactions between the PLA matrix and the MMT filler, which caused by the formation of polar interactions between the functional groups from both components. According to Liu et al. [16], the formation of hydrogen bonding between the Si–O groups of the MMT filler and the hydroxyl group of the PLA matrix was due to the peak's changes in the FTIR spectra.

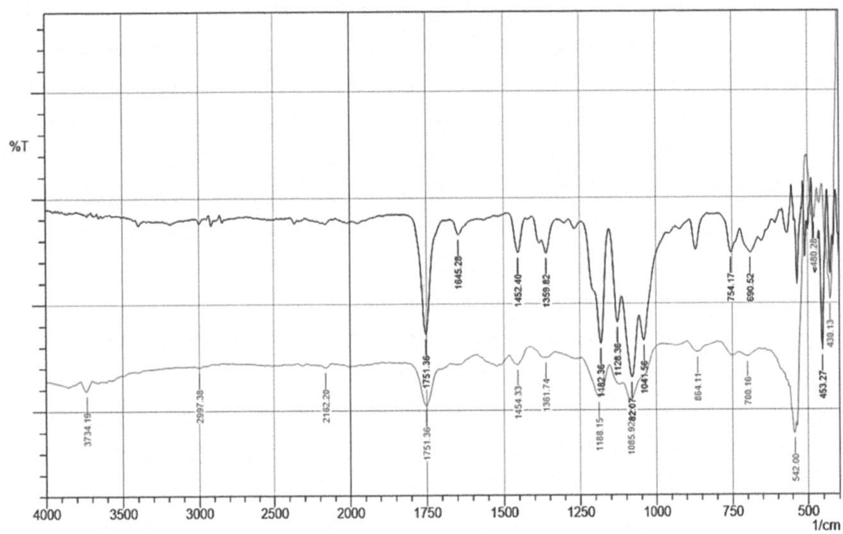

Fig. 11 FTIR spectra of (**a**) Pure PLA (Blue), and (**b**) Nanocomposite film (Green)

As observed in Fig. 11, the CF filler in the nanocomposite film did not influence the chemical structure of the nanocomposite, as there was no new functional group observed in the IR spectra of nanocomposite film. According to Arjmandi et al. [15], CF filler did not chemically interact with the PLA matrix, but only physical interaction occurred between them and this was confirmed by the SEM.

3.2 *Tensile Test Analysis*

Tensile testing was an analysis for mechanical properties of the nanocomposite that shown the effects of CF and MMT filler to the tensile strength and tensile modulus of the nanocomposite compared to the pure PLA. Tables 4 and 5 shows the formula used for tensile properties determination and the tensile properties of nanocomposite obtained from this research respectively.

Table 4 Determination of tensile properties

Tensile Strength (MPa)	$\frac{Load Applied(N)}{Crosssectionalarea(m^2)}$
Tensile Modulus (MPa)	$\frac{Stress(Pa)}{Strain}$

Table 5 Tensile strength and tensile modulus of nanocomposites film

No. samples	Tensile strength (MPa)	Tensile modulus (MPa)
1	1.304	51.581
2	0.650	28.526
3	0.691	51.027
4	2.442	84.778
5	1.450	113.251
6	1.315	85.279
7	0.632	38.551
8	1.608	34.478
9	1.596	50.893
10	1.477	38.145
11	1.242	48.147
12	1.348	45.683
13	0.906	67.982
14	0.663	17.371
15	1.454	23.093
PLA	0.931	17.592

3.2.1 Tensile Strength

Tensile strength of material was the ability of the material to withstand external force without breaking. Figure 12 shows the tensile strength of the nanocomposite with different formula composition design by the experiment design software Design Expert® version 11.

From the figure, it was noticed that the tensile strength of nanocomposite for Sample 4 was the highest at 2.442 MPa compared to the pure PLA. The design mixture for Sample 4 was observed to give the maximum tensile strength of the nanocomposite with 6.50wt% of bamboo cellulose fiber and 1.50wt% of MMT filler content. The tensile strength of nanocomposite for Sample 4 was approximately 62% higher than that of neat PLA, which suggesting a significant reinforcing effect to the nanocomposite. This finding led to the thought that the MMT filler content at this design was distributed uniformly within the PLA matrix, which contributes to the tensile strength to reaches its maximum value. This finding was supported by Arjmandi et al. [15], as similar observation on the effect of MMT on the PLA matrix. However, the tensile strength was significantly decreasing, as the MMT filler content rising exceeds 1.5wt%, which probably due to the aggregation of the filler that was resulting in a concentration point of fracture. This was confirmed by the SEM micrograph of nanocomposite.

In addition, it was noticed that there are few nanocomposites that possessed a lower tensile strength compared to the neat PLA film. This was due to the composition formulation of the nanocomposite, which may affect the mechanical strength of the nanocomposite. An excessive amount of cellulose fibers caused an obstruction that was attributed to the increased aggregation among cellulose fibers. This agreed

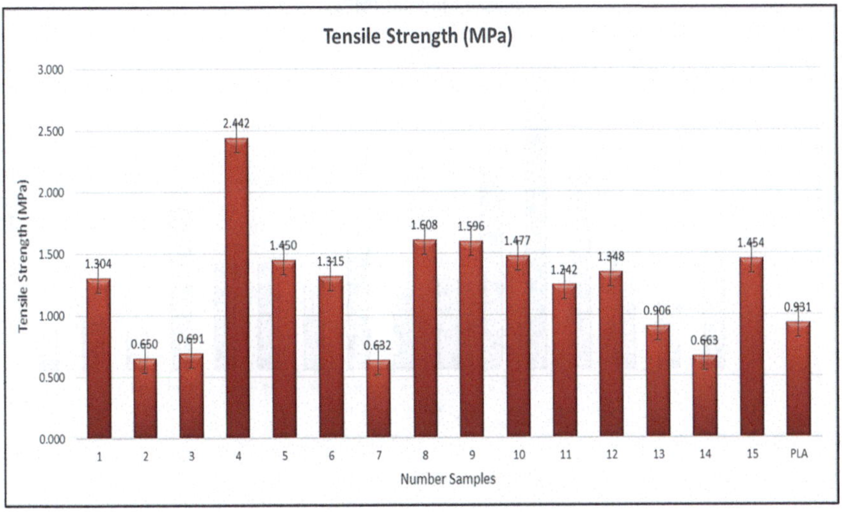

Fig. 12 Tensile strength (MPa) of the nanocomposites

with Arjmandi et al. [15] results, whereas the decrease in tensile strength of nanocomposite was attributed by the aggregation of the cellulose fibers in PLA matrix, which have been induced by the Van der Waals forces. The major factor influencing the trend of the tensile properties for the nanocomposite was the composition design obtained from the design experiment software, which gives the most optimum and least optimum design for the experiment.

3.2.2 Tensile Modulus

Generally, tensile modulus was measured by taking the amount of stress applied to the material and dividing it by the strain or elongation of the material undergoes. Tensile modulus was also known as elastic modulus, modulus of elasticity (MOE) and Young's modulus. Young' modulus also defined as the elongation of a material under a unit of stress. Below, Fig. 13 shown the tensile modulus obtained for all the samples of nanocomposite from tensile testing using a tensile testing machine.

From the figure, it was notable that the tensile modulus or MOE of nanocomposite film for Sample 5 (80wt% of PLA, 5wt% CF and 2wt% of MMT filler) increased by approximately 84% compared to the neat PLA film. According to Arjmandi et al. [15], this observation was attributed to the rigidity of MMT filler, which restricting the segmental chain motion in the polymer matrix. Figures 12 and 13 shows that almost all nanocomposites film increase in mechanical properties compared to the pure PLA film, which led to the observation that the adoption of CF and MMT filler into PLA matrix resulting in a positive reinforcing effect.

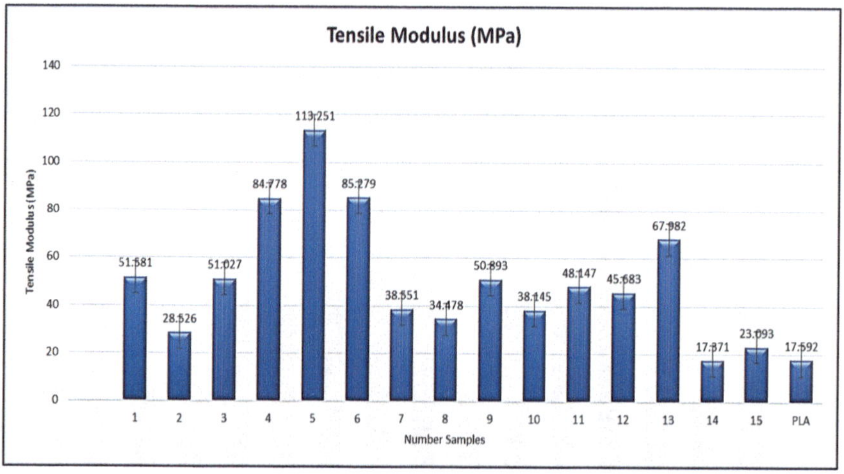

Fig. 13 Tensile modulus (MPa) of the nanocomposites

3.3 Morphological Analysis

It was necessary to study the morphology properties of the raw bamboo fibers, bamboo cellulose fibers and nanocomposite, since their mechanical properties are related to it. Figures 14 and 15 shows the micrograph of untreated GBF and CF, respectively. Figure 14 shows the morphology of the untreated bamboo fiber surface, which displayed no cavities and was covered by a layer of residues that responsible for fibrils compression. It was noticeable that the surface of untreated bamboo fiber was filled with starch grains, which become a layer that protect fibers and prevent the penetration of any resin into it. The material that was surfacing the bamboo fibrils was named cuticle and it was identified as aliphatic wax, which makes fibrils incompatible with most polymers. Thus, to make bamboo fibers suitable to be used as a reinforcing agent for PLA matrix or any other polymer, it needs to undergo chemical treatments beforehand.

The SEM micrograph surface of CF after underwent chemical treatments, such as dewaxing, delignification and pre-treatment. It was also called mercerization that helps to dissolved pectin and hemicellulose in the fibers and could increase the tensile strength of the fibers. The morphological surface of CF obtained illustrates a smooth appearance without having any micro void. It was clear that the CF displayed fibrils form shape and the chemical treatments had attributed to obvious differences in morphology form between CF and GBF. This micrograph confirms that mercerization resulting in fibrillation and the breakage of fiber into smaller pieces, which help in increased the effective surface area for contact with the polymer matrix. Similar findings were obtained by Liew et al. [14]. This finding was important to support the idea of enhancement PLA matrix by the implementation of bamboo cellulose fibers

Fig. 14 SEM micrograph of untreated raw green bamboo fiber (GBF)

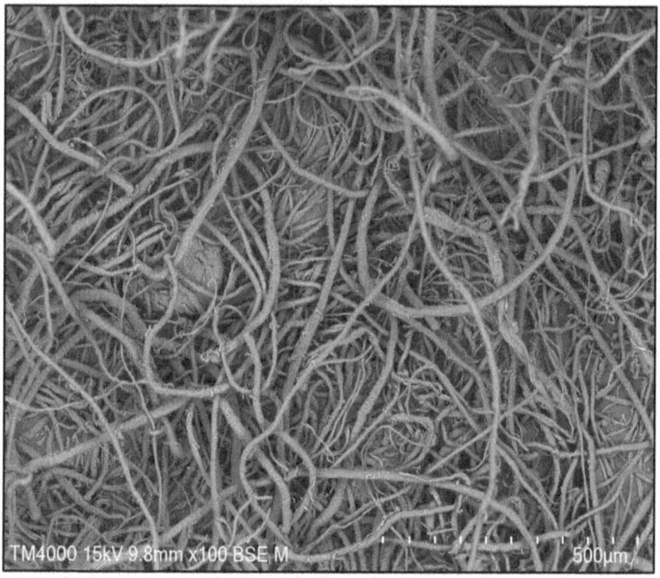

Fig. 15 SEM micrograph of bamboo cellulose fiber (CF)

and MMT filler into the matrix. This observation confirmed that there are obviously changes of properties of the bamboo fibers after underwent several treatments.

From the Fig. 16, the surface morphology of pure PLA film was typically flat and smooth as illustrated. Due to the amorphous characteristic of PLA matrix in all

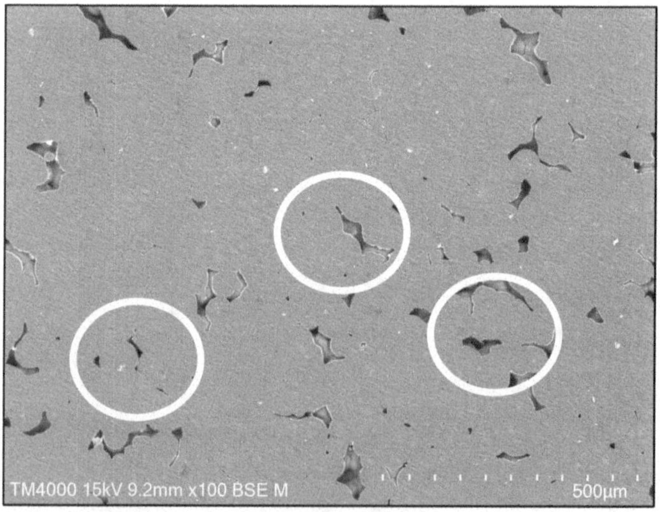

Fig. 16 SEM micrograph of pure PLA film

Fig. 17 SEM micrograph of nanocomposite film (Sample 4)

samples, the difference occurred in the matrix was studied, as the weak mechanical properties of PLA was the main reason for the enhancement of PLA matrix by adopting the cellulose fibers, and MMT filler into the matrix. As indicated by the small circles, it was clear that the pure PLA film displayed a lot of more micro voids compared to the surface of nanocomposite film that consisting PLA matrix, cellulose fibers and MMT filler.

From Fig. 17, the micrograph obtained for the nanocomposite film illustrates the fractured cross-sectional surface taken after underwent the tensile testing, which had exposed its micro-fibrillation structures. As indicated by the circles, there were some fibrillations destruction that occur from the tensile loading that showing the load was well transported through those cellulose fibers in the nanocomposite. The fractured cross-sectional surface showed that the cellulose fibers only physically interact with the PLA matrix and not chemically interact as confirmed by the chemical structure analysis of FT-IR.

Figure 18 above shows some of the aggregation of MMT filler at the surface of the nanocomposite film. This aggregation was observed to be the point of fracture for the nanocomposite film as the aggregated MMT filler possessed lower network formation of MMT filler-PLA matrix interactions which led to the decreasing of tensile strength of the nanocomposite. Similar observation was seen in the SEM micrograph of other nanocomposite films.

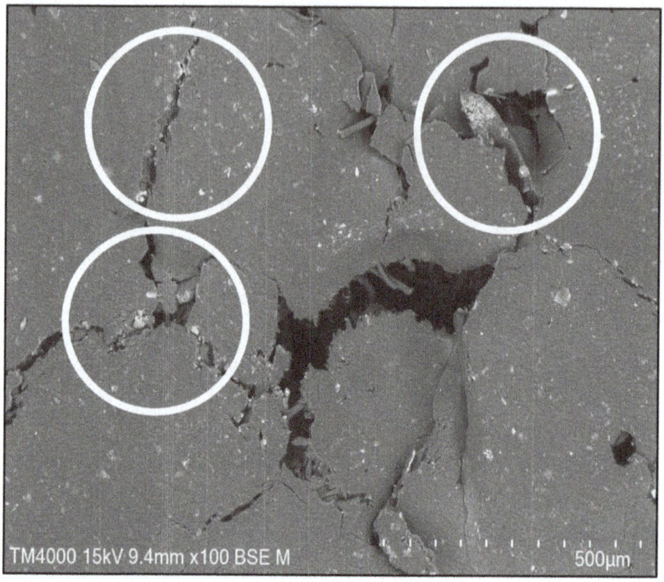

Fig. 18 SEM micrograph of nanocomposite film (Sample 5)

3.4 Energy Dispersive X-Ray Analysis (EDX/EDS)

Figure 19 shows the EDX/EDS spectrum of bamboo cellulose fibers. The spectrums peaks show that elements of carbon, oxygen and aluminum were detected from the EDX analysis. The elements of carbon and oxygen were originated from the bamboo cellulose fibers, while the aluminum element is from the aluminum holder for

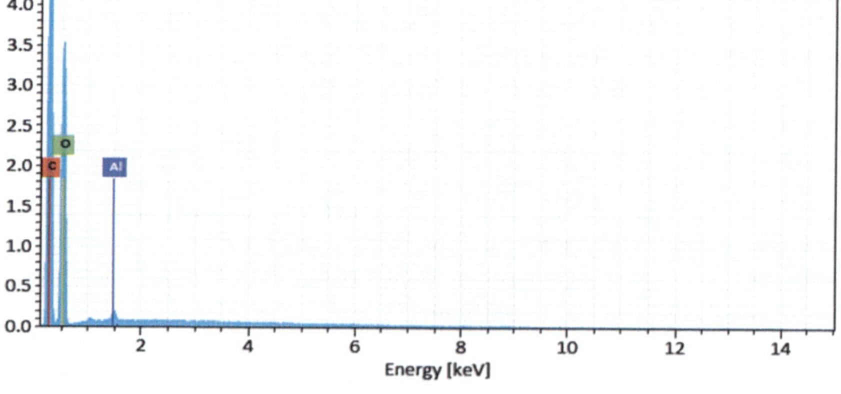

Fig. 19 EDX/EDS spectrum of bamboo cellulose fibers

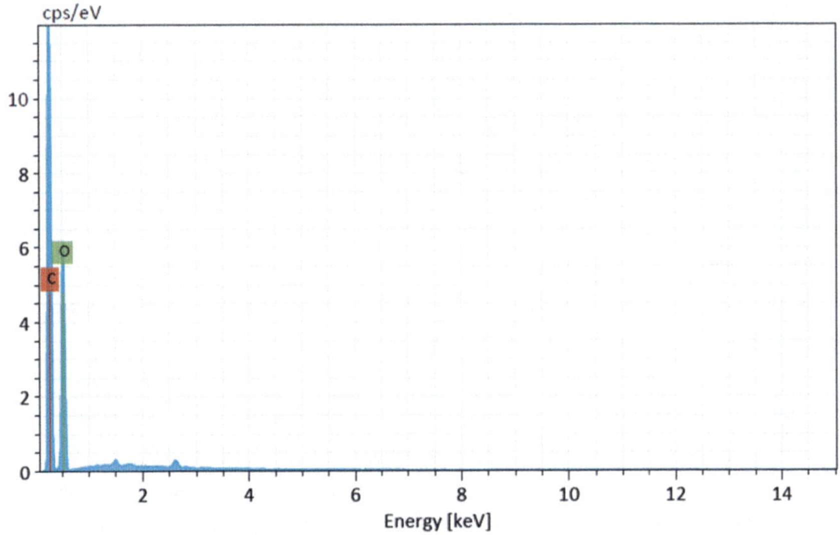

Fig. 20 EDX spectrum of pure PLA film

specimen used in the electron microscopy machine. Based on Fig. 19, the chemicals associated with the chemical's treatments were not detected in the final cellulose fibers. This observation was supported based on the analysis for the elemental analysis of cellulose fibers in the FT-IR spectra.

Figure 20 shows the EDX/EDS spectrum obtained from the pure PLA film. The main elements that detected in neat PLA were carbon and oxygen. This was due to the origin of PLA, which synthesis from the biomass or organic material. It was noticeable that the percentage of carbon higher compared to the percentage of oxygen in the material.

Figure 21 EDX/EDS spectrum for nanocomposite consisting the similar elements that was detected in the spectrum of bamboo cellulose fibers and pure PLA film, respectively. A new element detected in this spectrum which was element of Silicon. Silicon that appeared in this spectrum were originated from the MMT filler, which adopted into the PLA matrix along with the cellulose fibers. The similar functional group which consists silicon element also detected in the absorption peaks of FT-IR spectra of nanocomposite.

4 Conclusions

In conclusion, the bamboo cellulose fibers were successfully synthesized using the combination of both mechanical processes and chemical treatments. From this research, the FTIR analysis confirmed that the mercerization pre-treatment could

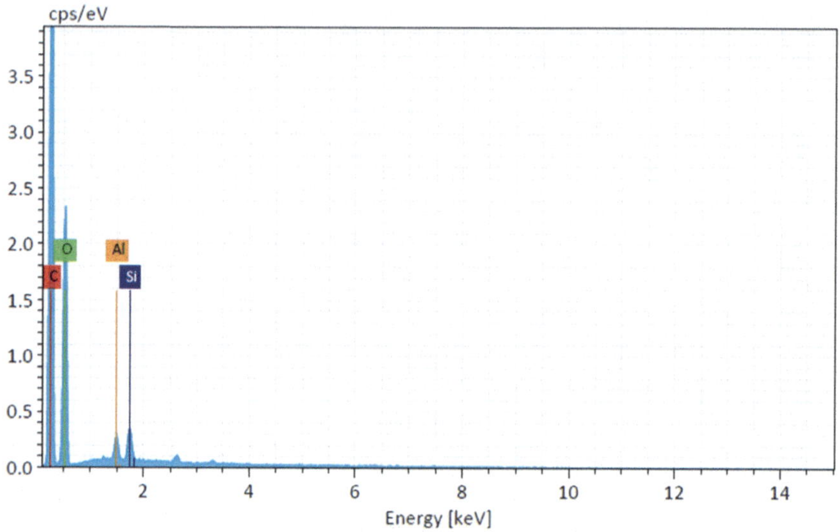

Fig. 21 EDX/EDS spectrum of nanocomposite film

remove the lignin and hemicellulose from the raw bamboo fibers. In addition, the nanocomposite consisting of PLA matrix, MMT filler and bamboo cellulose fibers were successfully prepared through solution casting technique. The solvents used for the solution casting technique was dichloromethane due to its safety and solubility properties. PLA nanocomposite reinforcement using bamboo cellulose fiber extracted from the raw green bamboo and with the association of MMT clay filler had shown improvement on the nanocomposite properties, whereas the mechanical properties of the nanocomposite for Sample 4 with 6.50wt% of bamboo cellulose fiber and 1.50wt% of MMT filler content were 62% higher compared to the pure PLA film. This improvement in the nanocomposite properties had a great potential to be implemented in the food packaging applications. Several testing and characterization techniques such as tensile testing, SEM analysis, FTIR analysis and etc. were done to study the effect of adopting the bamboo cellulose fibers and MMT clay filler into the PLA matrix. The mechanical properties of the nanocomposite were increased upon the addition of bamboo cellulose fibers and MMT clay filler compared to the neat PLA. The tensile strength and the tensile modulus (MOE) of the nanocomposite were improved by 62 and 79% respectively, for observation with Sample 4 that consists of 6.5wt% of bamboo cellulose fibers and 1.5wt% of MMT clay filler.

Acknowledgements The authors would like to acknowledge Universiti Malaysia Sarawak (UNIMAS) for the support.

References

1. Ramandhar, R., Kim, J.-H., Kim, J.-T.: Environmental, social and economic sustainability of bamboo and bamboo-based construction materials in buildings. J. Asian Arch. Build. Eng. **18**(2), 49–59 (2019). https://doi.org/10.1080/13467581.2019.1595629
2. Ingram, V., Tieguhong, J.C.: Bars to jars: bamboo value chains in cameroon. Ambio **42**(3), 320–333 (2013).https://doi.org/10.1007/s13280-012-0347-5
3. Yoo, J.-J., Divita, L., Kim, H.-Y.: Environmental awareness on bamboo product purchase intentions: do consumption values impact green consumption? Int. J. Fash. Des., Technol. Educ. **6**(1), 27–34 (2012). https://doi.org/10.1080/17543266.2012.758318
4. Bhonde, D., Nagarnaik, P. B., Parbat, D. K., Waghe, U. P.: Physical and mechanical properties of bamboo (Dendrocalmus strictus). Int. J. Sci. & Eng. Res. **5**(1), 455–459 (2014). https://www.ijser.org/researchpaper/Physical-and-Mechanical-Properties-of-Bamboo-Dendrocalmus-Strictus.pdf
5. Yasin, I., Haza, Z.F., Sutrisno, W.: Mechanical properties of bamboo as green materials to reduce the global warming effect. J. Adv. Res. Fluid. Mech. Therm. Sci .**52**(1), 46–54 (2018). https://www.akademiabaru.com/doc/ARFMTSV52_N1_P46_54.pdf
6. Chaowana, P.: Bamboo: an alternative raw material for wood and wood-based composites. J. Mater. Sci. Res. **2**(2), 90 (2013).https://doi.org/10.5539/jmsr.v2n2p90
7. Bubey, A., Jat, S.: A study on replacement of steel dowel bars by bamboo dowel bars. Int. Res. J. Eng. Technol. (IRJET) **7**(3), 179–182 (2020). https://www.irjet.net/archives/V7/i3/IRJET-V7I337.pdf
8. Lobovikov, M., Paudel, S., Piazza, M., Ren, H., Wu, J.: World bamboo resources: a thematic study prepared in the framework of the global forest resources assessment 2005 (No. 18). Rome (2007) https://www.fao.org/3/a1243e/a1243e00.htm
9. ASTM E168-16.: Standard practices for general techniques of infrared quantitative analysis, ASTM International, West Conshohocken, PA (2016). https://doi.org/10.1520/E0168-16
10. ASTM E1252-98.: Standard practice for general techniques for obtaining infrared spectra for qualitative analysis, ASTM International, West Conshohocken, PA (2013). https://doi.org/10.1520/E1252-98R13E01
11. ASTM E2015-04.: Standard guide for preparation of plastics and polymeric specimens for microstructural examination, ASTM International, West Conshohocken, PA (2014). https://doi.org/10.1520/E2015-04R14
12. ASTM E1508-12.: Standard guide for quantitative analysis by energy-dispersive spectroscopy, ASTM International, West Conshohocken, PA (2019). https://doi.org/10.1520/E1508-12AR19
13. ASTM D638-14.: Standard test method for tensile properties of plastics, ASTM International, West Conshohocken, PA (2014). https://doi.org/10.1520/D0638-14
14. Liew, F.K., Hamdan, S., Rahman, M., Rusop, M., Lai, J.C.H., Hossen, M.: Synthesis and characterization of cellulose from green bamboo by chemical treatment with mechanical process. J. Chem. **2015**(1), 212158 (2015). https://doi.org/10.1155/2015/212158
15. Arjmandi, R., Hassan, A., Haafiz, M. M., Zakaria, Z.: Effects of micro-and nano-cellulose on tensile and morphological properties of montmorillonite nanoclay reinforced polylactic acid nanocomposites. In Nanoclay reinforced polymer composites (pp. 103–125). Springer, Singapore (2016)
16. Liu, R., Luo, S., Cao, J., Peng, Y.: Characterization of organo-montmorillonite (OMMT) modified wood flour and properties of its composites with poly (lactic acid) Compos. A Appl. Sci. Manuf. **51**(33), 42 (2013)

Bamboo Nanocomposites Future Development and Applications

Md Rezaur Rahman, Perry Law Nyuk Khui, and Muhammad Khusairy Bin Bakri

Abstract Bamboo nanofibers, fibrils, and nanocomposites etc. has been notably researched throughout the years in the field of material science. Researchers demonstrate that bamboo possess good mechanical properties, biodegradability, rapidly growing properties, low costs, availability and sustainability. Bamboo nanocomposites are advantageous as a biodegradable material which could potentially replace numerous nondegradable conventional materials possessing similar properties and performance. Hence, this chapter summarizes the future development and applications of bamboo nanocomposites.

Keywords Bamboo · Nanocomposites · Sustainable · Development · Applications

1 Future Development of Bamboo Nanocomposites

Throughout the years, researchers have investigated different techniques to develop bamboo nanocomposites properties i.e. (mechanical, chemical, physical), including fabrication processes, utilizing design experimental models for response analysis and optimization, and varied material ratios of bamboo nanocomposites [9, 30, 37, 41–43]. The future development of bamboo nanocomposites may rely on the understanding of the advantages and disadvantages of bamboo followed by the optimization of bamboo nanocomposite fabrication parameters to cater for the desired performance and properties.

Examples of the advantages of bamboo includes;

- High mechanical strength and stiffness which are effective for load-bearing applications.
- Fast-growth (3–5 years), as a mature usable bamboo culm.
- Bamboo contains an abundance of high-quality cellulose fibers, useful for cellulose extraction and paper making.

M. R. Rahman (✉) · P. L. N. Khui · M. K. B. Bakri
Faculty of Engineering, Universiti Malaysia Sarawak, Jalan Datuk Mohammad Musa, 94300, Kota Samarahan, Sarawak, Malaysia
e-mail: rmrezaur@unimas.my

M. R. Rahman (ed.), *Bamboo Polymer Nanocomposites*, Engineering Materials,
https://doi.org/10.1007/978-3-030-68090-9_9

- Versatility of bamboo fibers ranges from the development woven fabric to reinforcing matrices of nanocomposites.
- Bamboo fiber has anti-bacterial properties.

Examples of the disadvantages of bamboo includes;

- Bamboo is prone to be invaded by wood-boring insects.
- It is recommended for manufacturers to store bamboo under dry conditions or treated with preservatives.
- Bamboo absorbs moisture and is sensitive to heat, light and ultraviolet radiation, if left untreated or stored under poor conditions.
- Bamboo cultivation is considered an invasive flora and spreads through roots and rhizomes.
- Difficulty of eradicating bamboo grove entirely.
- Bamboo fibers woven into fabrics are susceptible to shrinkage.

Despite the disadvantages stated, researchers have recurrently shown in advancing the development of bamboo's performance for many applications [3, 9, 37, 43]. However, researchers have found it difficult to maximize the economic and ecological strengths of bamboo. The bamboo industry has a low utilization rate of bamboo resources, which requires a lot of hands-on-work with little automation. Moreover, the bamboo industry produces limited products with no significant value in comparison to woods or timber where specific species are sought after [3].

Major challenges of developing bamboo nanocomposite include; the extraction of bamboo fibers as reinforcements, and quality control of fabricating bamboo nanocomposites. Researchers found it difficult to extract long, straight and fine bamboo fibers from the bamboo culm, utilizing the 3 main methods for extracting bamboo fibers [3, 37, 43].

1. Mechanical treatment i.e. (steam explosion method, retting, crushing, grinding, roll milling).
2. Chemical treatment i.e. (chemical retting, degumming, alkali or acid retting, surface modification).
3. Combination of mechanical and chemical treatments.

Bamboo culms are considered to be heterogeneous and anisotropic in nature, the extracted bamboo fibers mechanical properties are widely unstable due to its varied density and moisture content [36]. Researchers have reported bamboo culm's mechanical properties tend to improve as the bamboo plant matures. The improved mechanical properties of matured bamboo culms are associated to the increased specific gravity due to anatomical changes in the vascular bundles [36, 43].

Extraction process via. mechanical and chemical treatment methods most likely cause damage to the bamboo fibers, which usually results in producing shorter bamboo fiber lengths [1, 43]. However, the production of shorter bamboo fiber lengths is not a disadvantage, as majority of bamboo and wood-based nanocomposites are fabricated using particle reinforced composites methodologies. Bamboo nanofibers as reinforcements are normally dispersed randomly within the matrix, and mixed well to achieve homogeneity throughout the masterbatch [9, 18, 20, 30]. Thermoplastics and thermoset polymers are two types of polymer matrices used for developing bamboo nanocomposites. The compatibility, degradation rate, mechanical, thermal, and chemical properties of the polymer as some criteria used by researchers to select the matrix used for developing the nanocomposites. Moreover, at least one component in the fabrication process must utilize nanotechnology or characterized as nanoscale, in order to be considered as a nanocomposites [6, 32, 43].

The characterization of a single bamboo fiber includes the basic tests for morphology, chemical constituents and mechanical properties [3, 12]. The morphology of bamboo fiber presents a framework of multi-layered wall structure consist of concentric circles. The multi-layers consist of thick cell wall, small lumen, pits, and a small microfibril angle. The average dimension of bamboo fiber ranges from 10 to 30 mm in diameter and 1–4 mm in length [43]. The chemical constituents of bamboo are holocellulose 60–70%, pentosans 20–25%, hemicelluloses and lignin each amounted to about 20–30% and minor constituents such as resins, tannins, waxes and inorganic salts [12]. Khalil et al. reported similar chemical constituents of bamboo. The compiled database generally showed the major chemical constituents of bamboo are; cellulose, hemicelluloses and lignin, accounting for over 90% of the total mass, similar to wood-based sources [43]. Older research papers showed that α -cellulose content of bamboo is 40–50%, comparable with the ported α -cellulose contents of softwoods 40–52% and hardwoods 38–56% [10, 15, 31]. As higher cellulose content and aspect ratio contributes to increased stiffness to the nanocomposite, polymer matrices which exhibit ductile behaviours tend to become brittle and hard [15, 43]. The high lignin content of bamboo, which makes it a desirable construction material, contributes to its high heating value and its structural rigidity. Lignin also promotes reactivity, which makes for a smoother reaction to chemical change [10, 15, 43]. Moreover, the mechanical properties of bamboo fibers are dependent on numerous factors, such as; physical, chemical and morphological properties i.e. (fiber orientation, cellulose content, crystal structure, fiber dimensional length, diameter and cross-sectional area) [16, 43]. Different species of bamboo fibers yield different density and tensile strength, Defoirdt et al. found that the bamboo fiber species *Dendrocalamus membranaceus* has the density of 1.38 g/cm^3, and the tensile strength range between 638 to 813 MPa [34]. On the other hand, Monteiro et al. found that the bamboo fiber species *Bambusa vulgaris* has the density and tensile strength ranging between 1.03–1.21 g/cm^3 and 106–204 MPa respectively [17].

The optimization of bamboo nanocomposite is a difficult task such requires a lot of characterization tests via trial and error which is inefficient, time consuming and costly for researchers and manufacturers. Therefore, the future of bamboo nanocomposite fabrication parameters and testing may lie in use of simulations or design models to generate data and information, predicting outcomes based on initial experimental data. Examples of design of experiment model is D-optimal mixture design with surface response methodology, which can be use in a program by Stat-Ease "Design Expert" [4, 11, 41, 42]. Design of experiments is a statistical technique to evaluate linear, non-linear, and interactive effects of factors (independent variables) on responses (dependent variables) [2, 4, 21]. In comparison to other design of experiment techniques, D-optimal mixture design uses the sum of input variables to be one. In a mixture configuration, the measured response is therefore independent of the volume of the mixture and only depends on the relative proportions of components in the mixture [28]. Optimal modelling algorithms test unbiased process parameter estimates for the lowest number of experimental runs in a mixture design [5, 25].

An example of D-optimal mixture design is a study by M. Venkatesan et al., which optimizes the utilization of waste foundry sand and fly ash to synthesis geopolymer composite. The researcher uses D-optimal mixture design to study and process data for empirical model development, analysis of variance (ANOVA), diagnostics, and optimization of the geopolymer concrete compressive strength [4]. The model design generates iterations in weight percentage (wt%) of the independent variables; fine aggregates (A: 30–55 wt%), waste foundry sand (B: 15–40 wt%), and fly ash (C: 20–30 wt%), for the composite fabrication parameters. Experimental testing via compressive strength test (Y: N/mm^2) is the response and dependent variable set at 7th and 28th day curing time. In the D-optimal mixture design, the Scheffe cubic model was used, which is derived from standard polynomials of varying degrees to account for mixture variables [4, 38, 39]. M. Venkatesan et al. design the model which comprised of 14 test runs, including 6 runs for minimum model points, 4 runs for lack of fit estimation, and 4 runs for replicates.

Generally, the Scheffe cubic model represents the relationship between independent and dependent variables [4]. The optimization of the develop composite, correlates to the desirability percentage generated by the desirability function which combines all the optimization targets into one function. The desirability function maximises, as the factors and responses shift closer to the targets and become zero if either of the factors or reactions fall beyond their range of desirability. Results of optimization by all the factors and responses with highest desirability is selected. The researchers have noted the optimized composite weight percentage iteration are; 68 wt% fine aggregates, 38 wt% waste foundry sand and 30 wt% fly ash, with desirability of around 0.97. The optimized composite iteration can predict the compressive strength response to be within 18.9 N/mm^2 and 22.3 N/mm^2 for curing time of the 7th and 28th day respectively [4]. Researchers have shown the efficiency and practicality of using D-optimal mixture design to model to predict and optimize the composites fabrication parameters [4, 33, 39]. Optimization of composite fabrication parameters by design of experiments gives added advantage to numerous industries, by generating proper iterations for fabrication and optimize them based on the responses. The predictions calculated by the model are based around achieving the best fabrication parameters to optimize material usage in respect to the performance of composite responses, hence reducing manufacturing time and cost [4]. Fabrication parameter of bamboo nanocomposite is comparable to the design of experiment methodology by M. Venkatesan et al. using D-optimal mixture design, such as weight percentage of polymer matrix, bamboo nanofibers, curing time and processing temperature.

By isolating the bamboo plant to bamboo strips, culm and fibers, these materials are generally used to develop woven mats, veneer, and yarn, which are targeting the textile, interior design and furniture making industries. However, future trend of developing bamboo nanocomposite may lie in using cellulose nanofiber present in the bamboo. Extraction of cellulose nanofibers on other renewable source may yield different properties, i.e. (length, diameter, surface area, pore diameter, crystallinity, density and mechanical strength) [22, 27, 29, 30]. Lu et al. study the effects of chemical modification of bamboo cellulose nanofibers to improve mechanical properties of cellulose reinforced polylactic acid (PLA) composites [19]. The researchers use cellulose nanofiber extracted from bamboo and modifying with sodium hydroxide (NaOH) and silane coupling solution, which improves interfacial bonding with the PLA matrix. The study show that the mechanical properties of the bamboo cellulose nanofiber composite are dependent on the features of the cellulose nanofiber, matrix, and interfacial bonding. The disadvantages of the nanocomposite are the low compatibility and poor interfacial bonding between the hydrophilic cellulose nanofiber and the hydrophobic PLA matrix. Cellulose nanofiber has polar hydroxyl groups on the surface which form poor bonding interfaces with nonpolar PLA matrix, hence making it inefficient to distribute stresses under load and reducing the mechanical strength [19, 24]. The surface modification via chemical treatments enable the bamboo cellulose nanofiber composite to be applied in numerous industrial applications due to the improve mechanical properties which met industrial application requirements [19].

In general, the future development of bamboo nanocomposite is promising as it could be seen that researchers would include the use of the understanding of bamboo nanofiber advantages and disadvantages, design of experiment for setting up fabrication parameters to optimize testing, material usage, managing cost correlating to the nanocomposite performance. As well as exploring the usage of bamboo cellulose nanofiber for nanocomposite fabrication, catering for industrial applications.

2 Bamboo Nanocomposites Future Applications

The future application of bamboo nanocomposites is both dependent on the advancement of bamboo nanocomposite and the ability to cater towards a specific field of application. Over the years there are numerous advancement of bamboo nanocomposites, which uses bamboo fiber, additives and extracted bamboo cellulose nanofiber to modify properties of the fabricated nanocomposites. Researchers are consistently putting effort in improving the mechanical properties of bamboo nanocomposites and comparing it to other composites commercially available for various applications [7, 8, 13, 35].

Javadian et al. studies the design and application of sustainable bamboo fiber composite as reinforcements in structural concrete beams [26]. The development of the bamboo fiber composite reinforcement system characteristics such as mechanical properties, such as water absorption, swelling, shrinking, and chemical resistance were investigated by the researchers. The researchers found some advantages of using bamboo fibers over conventional synthetic fibers i.e. availability, renewability, biodegradability and lower production cost. The bamboo fibers are obtained by processing entire bamboo culms, then impregnate by using epoxy resin. The bamboo fiber composite is then molded via hot press fabrication method and divided into different sizes used in concrete beams in replacement of conventional steel bars [26]. As opposed to conventional building materials, bamboo fiber composite reinforcement shows low stiffness compared to conventional building materials. It is difficult to succeed in the mass construction industry. The material used for building application must meet specific criteria's and guidelines for mass construction, for bamboo composite reinforcements for building application. Structural deflections of buildings can be a critical factor for medium to high rise structures, where it would likely to develop cracks due to lower elastic modulus for the bamboo composite reinforcement. However, there are applications in which deflections are not as critical in the construction, such as low-rise low-cost housing units for developing countries. In this situation, deflection and the demand for ductile building material is low and secondary aspect failure would provide sufficient warning for an instability and collapsing structure [26]. Javadian et al. successfully demonstrate the potential of newly developed bamboo composites as a new type of reinforcement for non-deflection-critical applications of reinforced structural-concrete members [26]. Due to the bamboo fiber embedded in epoxy, the durability of the composite is greatly enhanced. In addition, flexural test results show the suitability according to ACI 440.1R-15 guidelines for the application of newly developed bamboo fiber composite reinforcement for structural concrete beams [26, 40].

By expanding into multiple disciplinaries, the future application of bamboo nanocomposite may gain greater potential and increase functionality. Bamboo cellulose nanofiber have gain interest by many researchers globally and advancement in the biomedical field. Kalia et al. noted advancement in cellulose nanofiber or nanocellulose in different aspects of biomedical developments include; scaffolds for tissue engineering, skin transplants for burns and wounds, drug release systems, artificial organs, blood vessel growth, stent covering, and bone reconstruction [23, 43].

An example of the use of bamboo cellulose nanofiber for biomedical application is by Oliveira et al. synthesizing bio-tissue based on nanocrystalline cellulose from Chilean bamboo *Chusquea quila* and cellulose acetate polymer solution matrix via electrospinning [14]. The researchers synthesize cellulose acetate bio-tissues with three iteration of bamboo nanocrystalline cellulose labelled BioT1, BioT5 and BioT10 which are 1%, 5% and 10% as a dry weight proportion of the acetate respectively. The morphological study shows good quality and more uniform diameter of the cellulose acetate nanofibers containing bamboo nanocellulose. A higher concentration of nanocellulose produced a more thermally unstable material that decomposed at lower temperatures. The BioT5 bio-tissue demonstrated the greatest resistance to traction with breakpoint of 30 MPa and an elasticity module of 1.597 MPa. The study by Oliveira et al. demonstrate the feasibility of developing cellulose nanostructured polymer fibers to reinforce their properties; expanding future applications in various fields ranging from construction to food packaging [14].

By applying bamboo fibers in various applications and multidisciplinary, new opportunities are open for researchers and manufacturers to design new and sustainable products. Cost-effective and eco-friendly nature of bamboo nanocomposites are some features which are advantageous for producing design new and sustainable products. In the near future, these composites are expected to see more and more uses. A detailed analysis of the fundamental, mechanical and physical properties of bamboo fibres is required to design such composites. Current research on bamboo fiber composites is focused on either the modification of the fiber or manipulating fiber weight percentages which influences the mechanical, thermal and other properties. However, the comprehensive usage of bamboo fiber and its nanocomposites are still far from ideal. Further research is required to address the challenges currently confronting the advanced application of bamboo fiber nanocomposites.

3 Summary

Researchers worldwide are working to developed nanocomposites using bamboo as the main reinforcement material with enhanced performance for global applications. In this chapter, the future development of bamboo nanocomposites may lie in the understanding of the advantages and disadvantages of bamboo, the use of design of experiment simulation models to optimize material usage, manage cost and fabrication parameters. Further research trends showed an affinity towards extracting and using nanocellulose from bamboo, as researchers attribute the influence of mechanical properties are related to higher cellulose content present in the bamboo. The future application of bamboo nanocomposite may lie in expanding into multidisciplinary such as biomedical industries, food and packaging, as well as mass construction. The application of nanocellulose derived from bamboo is also a trend in most applications in other fields not directly related to engineering.

Acknowledgements The authors would like to acknowledge Universiti Malaysia Sarawak (UNIMAS) for the support.

References

1. Zakikhani, P., Zahari, R., Sultan, M.T.H., Majid, D.L.: Extraction and preparation of bamboo fibre-reinforced composites. Mater. Des. **63**, 820–828 (2014)
2. Zaib, Q., Jouiad, M., Ahmad, F.: Ultrasonic synthesis of carbon nanotube-titanium dioxide composites: process optimization via response surface methodology. ACS Omega. **4**, 535–545 (2019)
3. Wang, G., Chen, F.: 10 - Development of bamboo fiber-based composites. In: Fan, M., F.B.T.-A.H.S.N.F.C. in Fu, C. (eds.), pp. 235–255. Woodhead Publishing (2017)
4. Venkatesan, M., Zaib, Q., Shah, I.H., Park, H.S.: Optimum utilization of waste foundry sand and fly ash for geopolymer concrete synthesis using D-optimal mixture design of experiments. Resour. Conserv. Recycl. **148**, 114–123 (2019)
5. Varanda, C., Portugal, I., Ribeiro, J., Silva, A., Silva, C.: Optimization of bitumen formulations using mixture design of experiments (MDOE). Constr. Build. Mater. **156**, 611–620 (2017)

6. Vallittu, P.K.: High-aspect ratio fillers: fiber-reinforced composites and their anisotropic properties. Dent. Mater. **31**, 1–7 (2015)
7. Thakur, A., Purohit, R., Rana, R.S., Bandhu, D.: Characterization and evaluation of mechanical behavior of epoxy-CNT-bamboo matrix hybrid composites. Mater. Today Proc. **5**, 3971–3980 (2018)
8. Tang, C.-M., Tian, Y.-H., Hsu, S.: Poly(vinyl alcohol) nanocomposites reinforced with bamboo charcoal nanoparticles: mineralization behavior and characterization. Materials (Basel) **8**, 4895–4911 (2015)
9. Sun, X., He, M., Li, Z.: Novel engineered wood and bamboo composites for structural applications: state-of-art of manufacturing technology and mechanical performance evaluation. Constr. Build. Mater. **249**, 118751 (2020)
10. Scurlock, J.M.O., Dayton, D., Hames, B.: Bamboo: an overlooked biomass resource? Biomass Bioenerg. **19**, 229–244 (2000)
11. Radfar, R., Hosseini, H., Farhoodi, M., Ghasemi, I., Średnicka-Tober, D., Shamloo, E., Mousavi Khaneghah, A.: Optimization of antibacterial and mechanical properties of an active LDPE/starch/nanoclay nanocomposite film incorporated with date palm seed extract using D-optimal mixture design approach. Int. J. Biol. Macromol. **158**, 790–799 (2020)
12. Prakash, C.: 7 - Bamboo fibre. In Kozłowski, R.M., M.B.T.-H. of N.F. (Second E. Mackiewicz-Talarczyk (eds.), Woodhead Publ. Ser. Text., pp. 219–229. Woodhead Publishing (2020)
13. Othman, M.H.: Bamboo fiber as fillers for polypropylene-nanoclay via injection molding. In Hashmi, S., I.A.B.T.-E. of R., Choudhury, S.M. (eds.), pp. 60–63. Elsevier, Oxford (2020)
14. Oliveira, P.E., Petit-Breuilh, X., Díaz, P.E., Gacitúa, W.: Manufacture of a bio-tissue based on nanocrystalline cellulose from chilean bamboo Chusquea quila and a polymer matrix using electrospinning. Nano-Struct. & Nano-Objects. **23**, 100525 (2020)
15. Mwaikambo, L.Y., Ansell, M.P.: The determination of porosity and cellulose content of plant fibers by density methods. J. Mater. Sci. Lett. **20**, 2095–2096 (2001)
16. Munawar, S., Umemura, K., Kawai, S.: Characterization of the morphological, physical, and mechanical properties of seven nonwood plant fiber bundles. J. Wood Sci. **53**, 108–113 (2007)
17. Monteiro, S.N., Lopes, F.P.D., Barbosa, A.P., Bevitori, A.B., Da Silva, I.L.A., Da Costa, L.L.: Natural lignocellulosic fibers as engineering materials—An overview. Metall. Mater. Trans. A **42**, 2963 (2011)
18. Mohanty, S., Nayak, S.: Short bamboo fiber-reinforced HDPE composites: influence of fiber content and modification on strength of the composite. J. Reinf. Plast. Compos. - J REINF PLAST Compos. **29**, 2199–2210 (2010)
19. Lu, T., Liu, S., Jiang, M., Xu, X., Wang, Y., Wang, Z., Gou, J., Hui, D., Zhou, Z.: Effects of modifications of bamboo cellulose fibers on the improved mechanical properties of cellulose reinforced poly(lactic acid) composites. Compos. Part B Eng. **62**, 191–197 (2014)
20. Llanos, J.H.R., Tadini, C.C.: Preparation and characterization of bio-nanocomposite films based on cassava starch or chitosan, reinforced with montmorillonite or bamboo nanofibers. Int. J. Biol. Macromol. **107**, 371–382 (2018)
21. Keinänen, P., Siljander, S., Koivula, M., Sethi, J., Sarlin, E., Vuorinen, J., Kanerva, M.: Optimized dispersion quality of aqueous carbon nanotube colloids as a function of sonochemical yield and surfactant/CNT ratio. Heliyon. **4**, e00787 (2018)
22. Kargarzadeh, H., Huang, J., Lin, N., Ahmad, I., Mariano, M., Dufresne, A., Thomas, S., Gałęski, A.: Recent developments in nanocellulose-based biodegradable polymers, thermoplastic polymers, and porous nanocomposites. Prog. Polym. Sci. **87**, 197–227 (2018)
23. Kalia, S., Dufresne, A., Cherian, B.M., Kaith, B.S., Avérous, L., Njuguna, J., Nassiopoulos, E.: Cellulose-Based Bio- and nanocomposites: a review. Int. J. Polym. Sci. **2011**, 837875 (2011)
24. Kalali, E.N., Hu, Y., Wang, X., Song, L., Xing, W.: Highly-aligned cellulose fibers reinforced epoxy composites derived from bulk natural bamboo. Ind. Crops Prod. **129**, 434–439 (2019)
25. Jeirani, Z., Mohamed Jan, B., Si Ali, B., Mohd. Noor, Chun Hwa, S., Saphanuchart, W.: The optimal mixture design of experiments: alternative method in optimizing the aqueous phase composition of a microemulsion, Chemom. Intell. Lab. Syst. **112**, 1–7 (2012)

26. Javadian, A., Smith, I.F.C., Hebel, D.E.: Application of sustainable bamboo-based composite reinforcement in structural-concrete beams: design and evaluation. Mater (Basel, Switzerland). **13**, 696 (2020)
27. Heinze, T., El Seoud, O.A., Koschella, A.: Production and characteristics of cellulose from different sources BT. In: Heinze, T., El Seoud, O.A., Koschella, A. (eds.) Cellulose Derivatives: Synthesis, Structure, and Properties, pp. 1–38. Springer International Publishing, Cham (2018)
28. Hare, L.: Experiments with mixtures: designs, models and the analysis of mixture data, 2nd ed. J. Qual. Technol. **23**, 168–169 (1991)
29. Han, S., Yao, Q., Jin, C., Fan, B., Zheng, H., Sun, Q.: Cellulose nanofibers from bamboo and their nanocomposites with polyvinyl alcohol: Preparation and characterization. Polym. Compos. **39**, 2611–2619 (2018)
30. Hai, L., Choi, E.S., Zhai, L., Panicker, P.S., Kim, J.: Green nanocomposite made with chitin and bamboo nanofibers and its mechanical, thermal and biodegradable properties for food packaging. Int. J. Biol. Macromol. **144**, 491–499 (2020)
31. Fengel, D., Shao, X.: A chemical and ultrastructural study of the bamboo species Phyllostachys makinoi Hay. Wood Sci. Technol. **18**, 103–112 (1984)
32. Faruk, O., Bledzki, A.K., Fink, H.-P., Sain, M.: Biocomposites reinforced with natural fibers: 2000–2010. Prog. Polym. Sci. **37**, 1552–1596 (2012)
33. Dilli Babu, G., Jagadish Babu, B., Bintu Sumanth, K., Sivaji Babu, K.: Experimental investigation on surface roughness of turned Nano-Khorasan based pineapple leaf fiber-reinforced polymer composites using response surface methodology. Mater. Today Proc. **27**, 2213–2217 (2020)
34. Defoirdt, N., Biswas, S., De Vriese, L., Tran, L.Q.N., Van Acker, J., Ahsan, Q., Gorbatikh, L., Van Vuure, A., Verpoest, I.: Assessment of the tensile properties of coir, bamboo and jute fibre. Compos. Part A Appl. Sci. Manuf. **41**, 588–595 (2010)
35. Chee, S.S., Jawaid, M., Sultan, M.T.H., Alothman, O.Y., Abdullah, L.C.: Effects of nanoclay on physical and dimensional stability of Bamboo/Kenaf/nanoclay reinforced epoxy hybrid nanocomposites. J. Mater. Res. Technol. **9**, 5871–5880 (2020)
36. Chaowana, P.: Bamboo: an alternative raw material for wood and wood-based composites. J. Mater. Sci. Res. **2** (2013)
37. Chand, N., Fahim, M.: 6 - Bamboo-reinforced polymer composites. In: Chand, N., M.B.T.-T. of N.F.P.C. (Second E. Fahim (eds.), Woodhead Publ. Ser. Compos. Sci. Eng., pp. 163–176. Woodhead Publishing (2021)
38. Bezener, M., Anderson, M., Whitcomb, P.: Formulation simplified: finding the sweet spot through design and analysis of experiments with mixtures (2018)
39. Arroyo-López, F.N., Bautista-Gallego, J., Chiesa, A., Durán-Quintana, M.C., Garrido-Fernández, A.: Use of a D-optimal mixture design to estimate the effects of diverse chloride salts on the growth parameters of Lactobacillus pentosus. Food Microbiol. **26**, 396–403 (2009)
40. American Concrete Institute, ACI 440.1R-15 Guide for the design and construction of structural concrete reinforced with fiber-reinforced polymer bars, ACI Committee 440, Farmington Hills, CA, USA, 2015
41. Adamu, M., Rahman, M.R., Hamdan, S., Bin Bakri, M.K., Yusof, F.A.B.M.: Impact of polyvinyl alcohol/acrylonitrile on bamboo nanocomposite and optimization of mechanical performance by response surface methodology. Constr. Build. Mater. **258**, 119693 (2020)
42. Adamu, M., Rahman, M.R., Hamdan, S.: Formulation optimization and characterization of bamboo/polyvinyl alcohol/clay nanocomposite by response surface methodology. Compos. Part B Eng. **176**, 107297 (2019)
43. Abdul Khalil, H.P.S., Alwani, M.S., Islam, M.N., Suhaily, S.S., Dungani, R., H'ng, Y.M., Jawaid, M.: 16 - The use of bamboo fibres as reinforcements in composites. In: Faruk, O., M.B.T.-B.R. in C.M. Sain (eds.), pp. 488–524. Woodhead Publishing (2015)

Educational and Awareness of Bamboo Nanocomposites Towards Sustainable Environment

Md Rezaur Rahman and Muhammad Khusairy Bin Bakri

Abstract This chapter discover the awareness of bamboo nanocomposites in the sustainable environment. Practically, bamboo nanocomposites can be applied as an alternative construction material. In order to employ in teaching and learning, it is necessary to uniform in the knowledge of prevention, mitigation, and adaptation strategies of nanocomposites materials. The agricultural and manufacturing-based models learning for bamboo, the models competency and standards models, future and current employability and educators build capacity also reported in this chapter.

Keywords Education · Awareness · Bamboo · Nanocomposites · Sustainable

1 Introduction

Education is important, so does its awareness. It is important to provide us with knowledge to understand things far better in many perspectives and views. Education also helps us to look deeper onto the meaning and purpose of certain life. It directly helps to build and create opinions, which subjected to a debate, whether it benefited the community or environment. Therefore, it is important to emphasize used of sustainable environment is very important in creating better living lifestyle and maintain the balance in the environment. According to Steenput [1], education involved wide coverage of awareness through presenting the knowledge, especially on the use of bamboo and its introduction topic. Theoretically and experimentally approached involve both lecturers and students, should be use in showing the existence of alternative construction materials, its material feels and its numerous construction possibilities [1].

M. R. Rahman (✉) · M. K. B. Bakri
Faculty of Engineering, Universiti Malaysia Sarawak, Jalan Datuk Mohammad Musa, 94300, Kota Samarahan, Sarawak, Malaysia
e-mail: rmrezaur@unimas.my

M. R. Rahman (ed.), *Bamboo Polymer Nanocomposites*, Engineering Materials,
https://doi.org/10.1007/978-3-030-68090-9_10

2 Education Implementations

In order to attract more people to learn more about the use of bamboo in such many applications, either its cover from non-conventional building materials or its design techniques, a proper model-based learning is required. According to Reid [2], the obsolete change in education systems was due to fast phase of change where human currently faced global crisis, either due to the human activities or natural disasters. Many researchers also stated that it is necessary to suit in the knowledge of prevention, mitigation, and adaptation strategies in education, especially involve in teaching and learning [3, 4].

3 Agricultural and Manufacturing-Based Models Learning for Bamboo

In agricultural and manufacturing-based models learning, there are five main considerations needed to be taken seriously: (i) the competency of the models, (ii) future and current employments, (iii) educators build capacity, and (iv) validations and skill gaps. Therefore, the implementations of those systems must be aligned and characterized to perform linearly with conformity and reasoned through the learning curriculum process, especially with the introduction of bamboo usage and design in manufacturing-based learning models. This system known to have little or less time and room to be tailored. In other words, easy to implement, change, harmonize and adopt with current new data and update in global based model learning. The model eco-system should be based on the use of the bamboo organic based sources, with a variety of other input where the sources permitted [5]. It also allowed feedbacks and opportunities to develop personal potential to its fullest, especially in both agricultural and manufacturing-based models for bamboo.

Due to the high demand of bamboo and its potentials in many applications, bamboo is widely used as a substitute to woods and other types of plastics and materials. This is directly related to those involves in structural and product applications, through improvements in processing technologies, product innovation, scientific and engineering skills [6]. Many countries had started to focus on bamboo-based products industry, which vastly potential for generating direct and side income and employment, especially those living in the rural and countryside areas [6]. Collaboration with several agencies in promoting the education and usage of bamboo had become another breaking point in creating new value-added modules and products. Institutions, i.e. universities and schools, especially lecturer and teacher should encourage students to become a better person, by building upon the passion of every student, and creating better conditions and environments for this to happen [7]. An improved agricultural model for our education system is really needed at the current moment to solve the global crisis. We need to create minds that can think differently and solved problems. Everyone must be able to think anew and excel in creating goo prospect of educations and environment. Therefore, everyone should be encouraged to learn and excel.

3.1 The Models Competency and Standards Models

There are two types of ways, which was useful to distinguish the model competency, there are: (i) competence structures models, and (ii) competence levels models. Several existence particular competence aspects were examined by the relationship relation to the overarching competence under study, whereas the latter describe various levels of competence differ qualitatively in terms of the task a person is able to perform given a particular level of competence [8]. In the competence levels models, which is related between people, the expectation of qualitative differences theoretical different levels was not always empirically tested [9]. Instead, to determine the score-based thresholds for levels, which is frequently defined post-hoc, test takers are assigned to a level based on their overall performance. However, the two model are not mutually exclusive.

According to Norris [10], there was is a close relationship between standards and competency. It is noted that some of the documents policy was read all competences are subject to standards, while, all standards referred to competence measures [11]. However, for this course, this does not have to be the case, as there are no specifics bamboo-based standards were set, other than its competences of pure knowledge, expenditure on universities, institutions, schools, i.e. students or pupils per class. The competences are skills, which will be taught and attained without being subjected to specific measurement and standard setting but indirectly related to it. Teaching was a standard-oriented learning process, each time students are presented with multiple prospects to acquire the compulsory competences through a cognitively stimulating and content-rich lesson design [9]. While it is not a case for 'standard-oriented', cognitively stimulating and content-rich lessons may well be desirable. Without the existence of such standards, the lessons were planned and taught perfectly well. Therefore, it was not clear such teaching is standard-oriented in any sense. Especially if anything was competence-oriented (even though it was not obviously the case) [9]. On the other hand, standards are not being described by the sentence quoted. The terms used as competence and standards interchangeable, thus makes it harder to discuss each concept properly regarding its meaning and implications [9].

As part of an educational output-oriented policy, standards are usually employed in education. The standard introduction justified the claimed of an improve teaching quality. As there was a danger that the standards existence results in a narrowing of both content and teaching method, it was not obvious that the standard introduction necessarily had this effect, and that teaching to the test occurs [9]. Such constitute developments cause teaching quality deterioration, which does not provide improvement. Thus, such policies merit provides careful discussion on the advantages and disadvantages, but more likely to be fruitful, as it is not conflated with what competences, and the way it was taught [9]. Likewise, the competences are questions, especially on are they acquired, and which ones are included in teaching, which are complex enough, and they are more likely to receive satisfactory answers [9]. The questions are examined separately based on the standards used that was set as the output-oriented context education policies. In an example, according to Pring [12, 13], whether someone is competent may be defined by referring to whether their tackling of situational demands meets the standard of having coped with the situation. However, by default, competence and standards do not belong together.

3.2 Future and Current Employability

According to Fugate et al. [14] and London [15, 16], the first employability step must relate to career motivation, which concerned around the direction, arousal, persistence and breadth of an individual's behavior and career-related decisions. Based on London [15], two factors involve multifaceted career motivation: (i) needs, and (ii) self-perceptions. This also includes the needs and drive for money, peer approval, security, advancement, and esteem [17]. In addition, the desire for continuous learning and generativity were also applied, i.e. passing knowledge on to the next generation [18]. An individual's self-perception to the realistic assessment of employment possibilities, as given the person perception of the relationship between the person's strengths and weaknesses and the career context, i.e. availability of jobs, organizational support for learning and development [14, 19, 20].

The second employability step is the human capital. Human capital includes the abilities, knowledge and skills of individuals gained through training, education, and job experiences [21]. A great research deal reported on the profound effect of education and competence development had earnings [22, 23], while others may argue that individuals as a human are those two variables, which is an important investment capital. To develop new competencies, an individual's willingness with positive attitude toward changing jobs and learning new skills for these jobs are also seen as important factors that enhance human capital [14, 24, 25].

Certain human capital aspects cannot be directly observed. Signaling theory suggests that some abilities, knowledge and skills are often assessed via signals [26–28]. The applicants may communicate the unobservable characteristic of quality, either by using years of training, schooling, job or organizational tenure history as signals of quality due to the lack of the employer's information about the job applications [27]. In addition, demographic characteristics, i.e. national origin, race, or disability status, also perceived as of human capital signals [25]. For example, employers may observe retirees as less attractive potential employees, i.e. lack of stamina, technological skills, and health [29]. Similarly, the employer's perceptions may contribute to the obstacle development that cause certain individuals' groups as their lower employability rate. For instance, due to recognized age discrimination, older workers perceived themselves as less likely to secure employment than younger workers [30].

The third step of employability is social capital, whereas the information., i.e. providing knowledge, advice, and emotional, i.e., listening, expressing concern, and support gained through other individual's relationship that used to identify and obtain career opportunities [31–34]. With diverse sets of people, including leaders, friends, family, community and business associates, i.e. coworkers, mentors, scholars have examined that the nature and individual's interaction quality influence the social capital [35, 36].

The identity is the fourth step of employability. Individual's self-definition of identity is an answer to the question of "Who am I?" [36]. Employability conceptualizations had an explicitly work domain that focused and described career identity as a degree, which a person defines themselves by their own work or career [14, 37]. Broader view to extend to the employability conceptualization is done by recognizing individuals had multiple identities and both work, i.e. professional, work group and organizer, and nonwork, i.e. child, friend, parent, and spouse identified their influence career decisions, options, and behaviors [36, 38, 39]. Many scholars also discussed the separate worlds myth that suggests both work and non-work identities are employability step, as it connected due to an individual's work and nonwork identities [40]. An individual's change in work or professional identity may influence their nonwork identity and vice versa. Ladge et al. [41] stated that pregnant women adapted to integrate a change in their nonwork identity (i.e., motherhood), the examined shows that the cross-domain identity transitions their established work identity. While others upon the interaction of their work and nonwork identities, a person examined their decisions making by change occupations or jobs. Ramarajan and Reid [40] detailed for telecommuters the blurring boundaries between the work and home domains, which cause them to renegotiate and reconstruct their nonwork and work identities. Therefore, the employability must reflect both nonwork and work identities.

The fifth employability step is personality. Several personality traits being linked to employability have been suggested [14, 42] being theoretically relevant to the construct with openness and proposed proactivity [14, 37, 43]. Openness usually refers as an individual's change acceptance and new experiences, as well as the willingness to enact change. In order to meet evolving work requirements High openness level person is flexible in uncertain situations and willing to take on new jobs, change tasks, or engage in learning [43, 44]. While proactivity refers which individuals extend to act, which affect their work context [45]. To identify career opportunities, proactive individuals expand their social networks and seek out information [37, 46].

3.3 Educators Build Capacity

For improved student learning in professional learning and enhanced teaching capacities in a range of areas, increased expectations of the institution and lecturer accountability comes the pressure, especially during teaching engagement [47–49]. Therefore, there should be no distinction made between the 'capacity building' and 'teacher professional learning' concepts, rather recognized that capacity building is a form and the component in teacher professional learning [50]. Althauser [51], Desimone [52], and Robinson et al. [52] investigated the professional learning initiatives and impacts on students and teachers. Althauser [51] asserts that to create positive change of professional development designed for student achievement, focus on the improve pedagogical and knowledge content, teach best practices, and readdress teachers teaching attitudes towards students is needed in learning requirements. While the capacity building part asserts that these are the elements of capacity building. This reflects that the goal established for the teachers in institutions to engage with capacity-building initiatives.

Penuel et al. [53] found that the effective professional development required both teachers provide technical support and time to plan for implementation. The professional job-embedded learning initiative was designed, to build teacher capacity to use student data provided teachers with a core skill [54]. The Capacity Building Interventions (CBIs) step ere developed by Marsh [55] and Marsh and Farrell [56] were evident in the implantation results. Their framework acknowledges four steps of variables impacted on the application and uptake of a CBI and on its effectiveness. These key dimensions are: (i) the nature of the interaction between players, (ii) focus practices of the CBI, (iii) artefacts employed in the CBI implementation and, (iv) the CBI contexts occurs.

The framework of Marsh [55] and Marsh and Farrell [56] assisted institutions in applying both sociocultural and social constructivist lens. It acts as an experiential tool to further develop theories about the lecturer, professional learning, CBIs and leadership in schools, especially in learning new subject related to bamboo. For understanding capacity building, these steps or variables provide a theoretical framework, while the only framing tool available to researchers. It was that best meet experiential of this research and the institution's context. Marsh [55] also recognized that a CBI support teacher in collecting and accessing data by organizing, filtering and analyzing it into a new form of information, either by combining the information with certain expertise. It also helped in adjust and respond to their instruction, by evaluating the outcome response effectiveness based on the result [55]. This also supports reflects set process demonstrated, thus affirming the work of Marsh and validating these findings. Marsh and Farrell [56] also developed a framework that further theorized the teacher building capacity in the data used. It also recognized the range of variables which impact the CBI effectiveness. Their framework also acknowledges these four steps or variables, which impacted the CBI application, uptake, and its effectiveness. These key dimensions are outlined, noting their application: (1) The interaction unit that the players interact with each other, i.e. group, one-on-one, novice-mentor. These interaction units included one-on-one and mentoring, small group dialogue, and whole-institutions conversations interactions. (2) The CBI practices used on the formality or informality of practices, i.e. observing, assessing individual teacher needs, modelling, sharing of experiences, provision of feedback such as coaching and mentoring, questioning, use of dialogue, and brokering with key players in the initiative. These practices collective roles was essential to the professional learning and capacity building of the teachers. They were shown in all evident in the school's data context. (3) The artefacts intervention employed could include symbols and tools that were created and adapted over time. The writing rubric, marking scale and its subsequent development over time was a noted artefact, which required to improve development. (4) The contexts in of intervention occur and includes in the cultural, historical and environmental facts, which become inherent in the process of building teachers' capacity with data used including institutions leadership [56]. The institutions multifarious factors evident in the culture and the chronological change initiatives, and development, whereas the physical and pedagogical environment were supportive and the capacity development.

3.4 Validations, Skills Gaps, Quality Teaching and Learning

For quality teaching and learning education, i.e. general, academic, technical and vocational, the main key features for the learners includes various strategies. This includes various supporting factors such as teacher and learner relationship and the learning environment method [57]. These strategies partly determined on the subject to be taught and naturalistic learners develops reciprocity and cooperation. The quality of teaching and learning strategy had been found to be suited in in vocational education and training, whereas it was identified as a respect enabling factor of good teaching [57]. Creating a culture is importance of for learners, whereas it encouraged a person to reach the inspirations and achieve which supported through their learning. Developing a purposeful and stimulating teaching environment was essential for effective teaching and broadening the skills gaps. This includes creating a classroom that was welcoming, which provides bright displays of learners' work that are informative. By fostering a sense of achievement, these environmental factors helped increase learner's motivation. For learning environments, it is identified that learning support is adapted to the needs of learners, as an important quality teaching and learning characteristic. Both on an individual and collective level, research is identified to be flexible and supportive, whilst at the same time they ensure challenges and attractive.

4 Example of Bamboo and Its Nanocomposites Learning Modules

4.1 Introduction to Bamboo

This module aims to introduce bamboo and its different types of species, demographics, contents, process, etc. This will prepare students to understand and determined each species strength, life duration, maturity, process, etc. The professional and experience agricultural industrial expert will assist the module program. Few article samples which can be used as references are shown in Table 1. This module

Table 1 Introduction to bamboo article samples

Authors	Title	References
Raj and Agarwal	Bamboo as a Building Materials	[58]
Abdul Khalil	Introductory Chapter: An Overview of Bamboo Research and Scientific Discoveries	[59]
Canavan et al.	The global distribution of bamboos: assessing correlates of introduction and invasion	[60]

required 2 credits time, which equivalent to 30 h learning time, which involve both theoretical and experimental aspects.

4.2 Bamboo and Its Nano Composites Processing and Fabrication Technique

Few modules are created aims to provide an information about bamboo extraction and process techniques. This includes different technique and process such as retting, pressing, chemical, etc. This module also includes few aspects of sustainability and renewability process. his module also aims to train and nurture students in learning the nanofabrication and treatment technique both simulation and experiment. The purpose of the module is to fulfill the shortage of potential talent and risk taker work force in in the industry in Malaysia, especially Sarawak. The module also covers the element of content creative design industry and the preparation of digital industrial. Few article samples which can be used as references are shown in Table 2. This module required 4 credits time, which equivalent to 60 h learning time, which involve both theoretical and experimental aspects.

4.3 Bamboo Plantations Techniques

This module aims to develop skills in formulating proper bamboo plantations planning, policy, standard of procedure and implementation. This also includes the soft skills and project management. The module also consists of designing and developing business model of both large and small-scale bamboo within international focus. Few article samples which can be used as references are shown in Table 3. This module required 2 credits time, which equivalent to 30 h learning time, which involve both theoretical and experimental aspects.

Table 2 Bamboo process article samples

Authors	Title	References
Huang et al.	Development of Bamboo scrimber: a literature review	[61]
Wang et al.	A systematic review on the composition, storage, processing of bamboo shoots: Focusing the nutritional and functional benefits	[62]
Adamu et al.	Bamboo nanocomposite: Impact of poly (ethylene-alt-maleic anhydride) and nanoclay on physicochemical, mechanical, and thermal properties	[63]
Hai et al.	Green nanocomposite made with chitin and bamboo nanofibers and its mechanical, thermal and biodegradable properties for food packaging	[64]
Han et al.	Cellulose nanofibers from bamboo and their nanocomposites with polyvinyl alcohol: Preparation and characterization	[65]
Guimares Junior et al.	Effect of the nano-fibrillation of bamboo pulp on the thermal, structural, mechanical and physical properties of nanocomposites based on starch/poly(vinyl alcohol) blend	[66]
Tang et al.	Poly(vinyl alcohol) Nanocomposites Reinforced with Bamboo Charcoal Nanoparticles: Mineralization Behavior and Characterization	[67]

Table 3 Bamboo plantations article samples

Authors	Title	References
Mera and Xu	Plantation management and bamboo resource economics in China	[68]
Francois and Liaw	Important Applications and the Perceived Benefits of Bamboo: A Comparison between Consumers and Businessmen	[69]

4.4 Bamboo Design and Structural Applications, Commercialization and Sustainability

This module aims to develop and train new design and structural applications for bamboo, its applications, commercialization and sustainability. The module also consists of designing a business model based on bamboo both large and small-scale within international focus. Few article samples which can be used as references are shown in Table 4. This module required 2 credits, which equivalent to 30 h learning time, which involve both theoretical and experimental aspects.

Table 4 Bamboo design and structural applications, commercialization and sustainability article samples

Authors	Title	References
Manandhar et al.	Environmental, social and economic sustainability of bamboo and bamboo-based construction materials in buildings	[70]

Acknowledgements The authors would like to acknowledge Universiti Malaysia Sarawak (UNIMAS) for the support.

References

1. Steenput, G.: Introducing bamboo in the education of the building engineer. Key Eng. Mater. **600**(1), 34–48 (2014). https://doi.org/10.4028/www.scientific.net/KEM.600.34
2. Reid, A.: Climate change education and research: possibilities and potentials versus problems and perils? Environ. Educ. Res. **25**(6), 767–790 (2019). https://doi.org/10.1080/13504622.2019.1664075
3. Berkhout, F., Hertin, J., Gann, D.M.: Learning to adapt: organisational adaptation to climate change impacts. Clim. Change **78**(1), 135–156 (2006). https://doi.org/10.1007/s10584-006-9089-3
4. Jickling, B.: Normalizing catastrophe: an educational response. Environ. Educ. Res. **19**(2), 161–176 (2013). https://doi.org/10.1080/13504622.2012.721114
5. Ben-zhi, Z., Mao-yi, F., Jin-zhong, X., Xiao-sheng, Y., Zheng-cai, L.: Ecological functions of bamboo forest: research and application. J. For. Res. **16**(1), 143–147 (2005). https://doi.org/10.1007/BF02857909
6. Gupta, A., Kumar, A.: Potential of bamboo in sustainable development. Asia-Pac. J. Manag. Res. Innov. **4**(3), 100–107 (2008). https://doi.org/10.1177/097324700800400312
7. Darling-Hammond, L., FLook, L., Cook-Harvey, C., Barron, B., Osher, D.: Implications for educational practice of the science of learning and development. Appl. Dev. Sci. **24**(2), 97–140 (2020). https://doi.org/10.1080/10888691.2018.1537791
8. Klieme, E., Hartiq, J., Rauch, D.: The concept of competence in educational contexts. In: Hartig, J., Klieme, E., Leutner, D. (eds.) Assessment of Competencies in Educational Contexts, pp. 3–22. Hogrefe & Huber Publishers (2008)
9. Glaesser, J.: Competence in educational theory and practice: a critical discussion. Oxford Rev. Educ. **1**(1), 70–85 (2018). https://doi.org/10.1080/03054985.2018.1493987
10. Norris, N.: The trouble with competence. Camb. J. Educ. **21**(3), 331–341 (1991). https://doi.org/10.1080/0305764910210307
11. Klieme, E., Avenrius, H., Blum, W., Dorbrih, P., Gruber, H., Prenxel, M., Resiss, K., Riquarts, K., Rost, J., Tenorth, H.-E., Vollmer, H.J.: Zur Entwicklung nationaler Bildungsstandards. Eine Expertise. Berlin: Bundesministerium für Bildung und Forschung (2007). https://sinus-transfer.uni-bayreuth.de/fileadmin/MaterialienBT/Expertise_Bildungsstandards.pdf
12. Pring, R.: Standards and quality in education. Br. J. Educ. Stud. **40**(1), 4–22 (1992). https://doi.org/10.1080/00071005.1992.9973907
13. Pring, R.: The skills revolution. Oxford Rev. Educ. **30**(1), 105–116 (2004). https://doi.org/10.1080/0305498042000190078
14. Fugate, M., Kinichi, A.J., Ashforth, B.E.: Employability: a psycho-social construct, its dimensions, and applications. J. Vocat. Behav. **65**(1), 14–38 (2004). https://doi.org/10.1016/j.jvb.2003.10.005

15. London, M.: Toward a theory of career motivation. Acad. Manag. Rev. **8**(4), 620–630 (1983). https://doi.org/10.5465/amr.1983.4284664
16. London, M.: Relationships between career motivation, empowerment and support for career development. J. Occup. Organ. Psychol. **66**(1), 55–69 (1993). https://doi.org/10.1111/j.2044-8325.1993.tb00516.x
17. Alderfer, C.P.: An empirical test of a new theory of human needs. Organ. Behav. Hum. Perform. **4**(2), 142–175 (1969). https://doi.org/10.1016/0030-5073(69)90004-X
18. McAdams, D.P., de St. Aubin, E., Logan, R.L.: Generativity among young, midlife, and older adults. Psychol. Aging **8**(2), 221–230 (1993). https://doi.org/10.1037/0882-7974.8.2.221
19. Thijssen, J.G., Van der Heijden, B.I., Rocco, T.S.: Toward the employability—link model: current employment transition to future employment perspectives. Hum. Resour. Dev. Rev. **7**(2), 165–183 (2008). https://doi.org/10.1177/1534484308314955
20. Vanhercke, D., De Cuyper, N., Peeters, E., De Witte, H.: Defining perceived employability: a psychological approach. Pers. Rev. **43**(4), 592–605 (2014). https://doi.org/10.1108/PR-07-2012-0110
21. Sullivan, S.E., Ariss, A.A.: Employment after retirement: a review and framework for future research. J. Manag. **45**(1), 262–284 (2018). https://doi.org/10.1177/0149206318810411
22. Judge, T.A., Cable, D.M., Boudreau, J.W., Bretz, R.D.: An empirical investigation of the predictors of executive career success. Pers. Psychol. **48**(3), 485–519 (1995). https://doi.org/10.1111/j.1744-6570.1995.tb01767.x
23. Judge, T.A., Higgins, C.A., Thoresen, C.J., Barrick, M.R.: The big five personality traits, general mental ability, and career success across the life span. Pers. Psychol. **52**(3), 621–652 (1999). https://doi.org/10.1111/j.1744-6570.1999.tb00174.x
24. Van der Heijde, C.M., van der Heijden, B.: A competence-based and multidimensional operationalization and measurement of employability. Hum. Resour. Manag. J. **45**(3), 449–476 (2006). https://doi.org/10.1002/hrm.20119
25. Wittekind, A., Raeder, S., Grote, G.: A longitudinal study of determinants of perceived employability. J. Organ. Behav. **31**(4), 566–586 (2010). https://doi.org/10.1002/job.646
26. Connelly, B.L., Certo, S.T., Ireland, R.D., Reutzel, C.R.: Signaling theory: a review and assessment. J. Manag. **37**(1), 39–67 (2011). https://doi.org/10.1177/0149206310388419
27. Karasek, R., Bryant, P.: Signaling theory: past, present, and future. Acad. Strat. Manag. J. **11**(1), 91–99 (2012)
28. Spence, M.: Job market signaling. Quart. J. Econ. **87**(3), 355–374 (1973). https://doi.org/10.2307/1882010
29. Patrickson, M., Ranzijn, R.: Employability of older workers. Equal. Oppor. Int. **22**(5), 50–63 (2003). https://doi.org/10.1108/02610150310787496
30. Kovalenko, M., Mortelmans, D.: Contextualizing employability: do boundaries of self-directedness vary in different labor market groups? Career Dev. Int. **21**(5), 498–517 (2016). https://doi.org/10.1108/CDI-01-2016-0012
31. Adler, P.S., Kwon, S.W.: Social capital: prospects for a new concept. Acad. Manag. Rev. **27**(1), 17–40 (2002). https://doi.org/10.2307/4134367
32. Forret, M.L. (2018). Networking as a job search and career management behavior. In: Klehe, U.C., Van Hooft, E.W.J. (eds.) The Oxford Handbook of Job Loss and Job Search, pp. 275–291. Oxford University Press, New York. https://doi.org/10.1093/oxfordhb/9780199764921.013.022
33. McArdle, S., Waters, L., Briscoe, J.P., Hall, D.T.T.: Employability during unemployment: adaptability, career identity and human and social capital. J. Vocat. Behav. **71**(2), 247–264 (2007). https://doi.org/10.1016/j.jvb.2007.06.003
34. Forret, M.L., Dougherty, T.W.: Correlates of networking behavior for managerial and professional employees. Group & Organ. Manag. **26**(3), 283–311 (2001). https://doi.org/10.1177/1059601101263004
35. Lin, N., Ensel, W.M., Vaughn, J.C.: Social resources and strength of ties: structural factors in occupational status attainment. Am. Sociol. Rev. **46**(4), 393–405 (1981). https://doi.org/10.2307/2095260

36. Miscenko, D., Day, D.V.: Identity and identification at work. Organ. Psychol. Rev. **6**(3), 215–247 (2016). https://doi.org/10.1177/2041386615584009
37. Fugate, M., Kinicki, A.J.: A dispositional approach to employability: development of a measure and test of implications for employee reactions to organizational change. J. Occup. Organ. Psychol. **81**(3), 503–527 (2008). https://doi.org/10.1348/096317907X241579
38. Ramarajan, L.: Past, present and future research on multiple identities: toward an intrapersonal network approach. Acad. Manag. Ann. **8**(1), 589–659 (2014). https://doi.org/10.1080/19416520.2014.9123
39. Welbourne, T.M., Paterson, T.A.: Advancing a richer view of identity at work: the role-based identity scale. Pers. Psychol. **70**(2), 315–356 (2016). https://doi.org/10.1111/peps.12150
40. Ramarajan, L., Reid, E.: Shattering the myth of separate worlds: negotiating nonwork identities at work. Acad. Manag. Rev. **38**(4), 621–644 (2013). https://doi.org/10.5465/amr.2011.0314
41. Ladge, J.J., Clair, J.A., Greenberg, D.: Cross-domain identity transition during liminal periods: constructing multiple selves as professional and mother during pregnancy. Acad. Manag. J. **55**(6), 1449–1471 (2012). https://doi.org/10.5465/amj.2010.0538
42. Bal, P.M., De Jong, S.B., Jansen, P.G., Bakker, A.B.: Motivating employees to work beyond retirement: a multi-level study of the role of I-deals and unit climate. J. Manage. Stud. **49**(2), 306–331 (2012). https://doi.org/10.1111/j.1467-6486.2011.01026.x
43. Nauta, A., Vianen, A., Heijden, B., Dam, K., Willemsen, M.: Understanding the factors that promote employability orientation: the impact of employability culture, career satisfaction, and role breadth self-efficacy. J. Occup. Organ. Psychol. **82**(2), 233–251 (2009). https://doi.org/10.1348/096317908X320147
44. Caspi, A., Roberts, B.W., Shiner, R.L.: Personality development: stability and change. Annu. Rev. Psychol. **56**(1), 453–484 (2005). https://doi.org/10.1146/annurev.psych.55.090902.141913
45. Bateman, T.S., Crant, J.M.: The proactive component of organizational behavior: A measure and correlates. J. Organ. Behav. **14**(2), 103–118 (1993). https://doi.org/10.1002/job.4030140202
46. Trusty, J., Allen, D.G., Fabian, F.: Hunting while working: an expanded model of employed job search. Hum. Resour. Manag. Rev. (2018). https://doi.org/10.1016/j.hrmr.2017.12.001
47. Klenowski, V.: Assessment for learning in the accountability era: Queensland Australia. Stud. Educ. Eval. **37**(1), 78–83 (2011). https://doi.org/10.1016/j.stueduc.2011.03.003
48. Nadelson, L.S., Fuller, M., Briggs, P., Hammons, D., Bubak, K., Sass, M.: The tension between teacher accountability and flexibility: the paradox of standards-based reform. Teach. Educ. Pract. **25**(2), 196–220 (2012)
49. Thompson, G.: NAPLAN, MySchool and accountability: teacher perceptions of the effects of testing. Int. Educ. J.: Comp. Perspect. **12**(2), 62–84 (2014). https://files.eric.ed.gov/fulltext/EJ1017709.pdf
50. Johnston, J., Goerge, S.: A tool for capacity building: teacher professional learning about teaching writing. Teacher Dev. **22**(5), 685–702 (2018). https://doi.org/10.1080/13664530.2018.1484389
51. Althauser, K.: Job-embedded professional development: its impact on teacher self-efficacy and student performance. Teacher Dev. **19**(2), 210–225 (2015). https://doi.org/10.1080/13664530.2015.1011346
52. Robinson, J., Myran, S., Strauss, R., Reed, W.: The impact of an alternative professional development model on teacher practices in formative assessment and student learning. Teacher Dev. **18**(2), 141–162 (2014). https://doi.org/10.1080/13664530.2014.900516
53. Penuel, W.R., Fishman, B.J., Yamaguchi, R., Gallagher, L.P.: What makes professional development effective? Strategies that foster curriculum implementation. Am. Educ. Res. J. **44**(4), 921–958 (2007). https://doi.org/10.3102/0002831207308221
54. Lynch, D., Smith, R., Provost, S., Madden, J.: Improving teaching capacity to increase student achievement: the key role of data interpretation by school leaders. J. Educ. Adm. **54**(5), 575–592 (2016). https://doi.org/10.1108/JEA-10-2015-0092

55. Marsh, J.A.: Interventions promoting educators' use of data: research insights and gaps. Teach. Coll. Rec. **114**(11), 1–48 (2012)
56. Marsh, J.A., Farrell, C.C.: How leaders can support teachers with data-driven decision making: a framework for understanding capacity building. Educ. Manag. Adm. Leadersh. **43**(2), 269–289 (2015). https://doi.org/10.1177/1741143214537229
57. Mishra, P.T., Mishra, A., Chowhan, S.S.: Role of higher education in bridging the skill gap. Univers. J. Manag. **7**(4), 134–139 (2019). https://doi.org/10.13189/ujm.2019.070402
58. Raj, D.A., Agarwal, B.: Bamboo as a building material. J. Civ. Eng. Environ. Technol. **1**(3), 56–61 (2014). https://www.krishisanskriti.org/vol_image/03Jul201502074415.pdf
59. Abdul Khalil, H.P.S.: Introductory chapter: an overview of bamboo research and. In: Abdul Khalil, H.P.S. (ed.) Bamboo—Current and Future Prospects, pp. 3–6, IntechOpen, London (2018). https://doi.org/10.5772/intechopen.77094
60. Canavan, S., Richardon, D.M., Visser, V., Le Roux, J.J., Vorontsova, M.S., Wilson, J.R.U.: The global distribution of bamboos: assessing correlates of introduction and invasion. AoB Plants **9**(1), plw079 (2017). https://doi.org/10.1093/aobpla/plw078
61. Huang, Y., Ji, Y., Yu, W.: Development of bamboo scrimber: a literature review. J. Wood Sci. **65**(1), 25 (2019). https://doi.org/10.1186/s10086-019-1806-4
62. Wang, Y., Chen, J., Wang, D., Ye, F., He, Y., Hu, Z., Zhao, G.: A systematic review on the composition, storage, processing of bamboo shoots: focusing the nutritional and functional benefits. J. Funct. Foods **71**(1), 104015 (2020). https://doi.org/10.1016/j.jff.2020.104015
63. Adamu, M., Rahman, M.R., Hamdan, S.: Bamboo nanocomposite: impact of poly (ethylene-alt-maleic anhydride) and nanoclay on physicochemical, mechanical, and thermal properties. BioResources **15**(1), 331–346 (2019). https://bioresources.cnr.ncsu.edu/resources/bamboo-nanocomposite-impact-of-poly-ethylene-alt-maleic-anhydride-and-nanoclay-on-physicochemical-mechanical-and-thermal-properties/
64. Hai, L., Choi, E.S., Zhai, L., Panicker, P.S., Kim, J.: Green nanocomposite made with chitin and bamboo nanofibers and its mechanical, thermal and biodegradable properties for food packaging. Int. J. Biol. Macromol. **144**(1), 491–499 (2020). https://doi.org/10.1016/j.ijbiomac.2019.12.124
65. Han, S., Yao, Q., Jin, C., Fan, B., Zheng, H., Sun, Q.: Cellulose nanofibers from bamboo and their nanocomposites with polyvinyl alcohol: preparation and characterization. Polym. Compos. **39**(8), 2611–2619 (2016). https://doi.org/10.1002/pc.24249
66. Guimarães Junior, M., Teixeira, F.G., Tonoli, G.H.D.: Effect of the nano-fibrillation of bamboo pulp on the thermal, structural, mechanical and physical properties of nanocomposites based on starch/poly(vinyl alcohol) blend. Cellulose **25**(1), 1823–1849 (2018). https://doi.org/10.1007/s10570-018-1691-9
67. Tang, C.-M., Tian, Y.-H., Hsu, S.-H.: Poly(vinyl alcohol) nanocomposites reinforced with bamboo charcoal nanoparticles: mineralization behavior and characterization. Materials **8**(8), 4895–4911 (2015). https://doi.org/10.3390/ma8084895
68. Mera, F.A.T., Xu, C. (2014) Plantation management and bamboo resource economics in china. Revista Ciencia Y Technologia **7**(1), 1–2 (2014). https://doi.org/10.18779/cyt.v7i1.137
69. Francois, D., Liaw, S.-Y.: Important applications and the perceived benefits of bamboo: a comparison between consumers and businessmen. Int. J. Bus. Manag. **14**(6), 12–28 (2019). https://doi.org/10.5539/ijbm.v14n6p12
70. Manandhar, R., Kim, J.-H., Kim, J.T.: Environmental, social and economic sustainability of bamboo and bamboo-based construction materials in buildings. J. Asian Arch. Build. Eng. **18**(2), 49–59 (2019). https://doi.org/10.1080/13467581.2019.1595629

The manufacturer's authorised representative in the EU is Springer Nature Customer Service Centre GmbH, Europaplatz 3, 69115 Heidelberg, Germany. If you have any concerns regarding our products, please contact ProductSafety@springernature.com

Printed and bound by CPI Group (UK) Ltd, Croydon, CR0 4YY
22/09/2026
02221055-0002